Understanding Perspectivism

This edited collection is the first of its kind to explore the view called perspectivism in the philosophy of science. The book brings together an array of essays that reflect on the methodological promises and scientific challenges of perspectivism in a variety of fields such as physics, biology, cognitive neuroscience, and cancer research, just for a few examples. What are the advantages of using a plurality of perspectives in a given scientific field and for interdisciplinary research? Can different perspectives be integrated? What is the relation between perspectivism, pluralism, and pragmatism? These ten new essays by top scholars in the field offer a kaleidoscopic journey toward understanding the view called "perspectivism" and its relevance to science.

Michela Massimi is Professor of Philosophy of Science at the University of Edinburgh, UK. She was Co-editor-in-Chief of the *British Journal for the Philosophy of Science* (2011–2016) and Vice President of the *European Philosophy of Science Association* (2015–2019). She is the Principal Investigator of the ERC-funded project "Perspectival Realism. Science, knowledge and truth from a human vantage point" (2016–2020).

Casey D. McCoy is a Postdoc at Stockholm University, Sweden. His research falls primarily within the philosophy of science and the philosophy of physics, and he has written on topics including inflationary cosmology, fine-tuning problems in physics, and the interpretation of statistical mechanics.

Routledge Studies in the Philosophy of Science

For more information about this series, please visit: www.routledge.com/Routledge-Studies-in-the-Philosophy-of-Science/book-series/POS

This impressive collection is essential reading for appreciating the inevitable contextualities of scientific knowledge. It explores how notions of "perspective" can illuminate the epistemic upshot of the sciences and how they are situated in their history, practices, representations, and sometimes competing aims, provocatively advancing debates about realism, pragmatism, explanation, and modeling in the process, all through a wealth of cases from physics, biology, neuroscience, and medical science.

—**Anjan Chakravartty**, University of Miami

An excellent collection of essays on a topic rapidly establishing itself as an important interpretive programme in philosophy of science. One of the volume's many merits consists in showing the diversity and versatility of perspectivism while illustrating common features among its different varieties. The reader is thus provided an enormously rich foundation for evaluating the role of perspectivism in understanding science and its practices.

—**Margaret Morrison**, University of Toronto

Perspectivism is a fruitful metaphor for imagining alternatives to traditional realism in philosophy of science. Massimi and McCoy have gathered ten essays which show how perspectivism is illuminating in areas such as molecular biology and measurement theory, and also explore the relationships between perspectivism and other recent accounts including pragmatism, structural realism, pluralism, and scientific modelling. There is an excellent balance of established and emerging scholars in the field. This volume is a superb, cutting-edge text to use in an advanced graduate seminar.

—**Miriam Solomon**, Temple University

Understanding Perspectivism

Scientific Challenges and
Methodological Prospects

**Edited by Michela Massimi and
Casey D. McCoy**

Routledge
Taylor & Francis Group
NEW YORK AND LONDON

First published 2020
by Routledge
605 Third Avenue, New York, NY 10017

and by Routledge
2 Park Square, Milton Park, Abingdon, Oxon, OX14 4RN

First issued in paperback 2021

Routledge is an imprint of the Taylor & Francis Group, an informa business

Publisher's Note
The publisher has gone to great lengths to ensure the quality of
this reprint but points out that some imperfections in the
original copies may be apparent.

Library of Congress Cataloging-in-Publication Data
Names: Massimi, Michela, editor. | McCoy, Casey D.
   (Casey David), editor.
Title: Understanding perspectivism : scientific and methodological
   prospects / edited by Michela Massimi and Casey D. McCoy.
Description: New York : Taylor & Francis, 2019. | Series: Routledge
   studies in the philosophy of science ; 20 | Includes bibliographical
   references and index.
Identifiers: LCCN 2019006700 | ISBN 9781138503069 (hardback)
Subjects: LCSH: Science—Philosophy. | Opinion (Philosophy)
Classification: LCC Q175 .U475 2019 | DDC 501—dc23
LC record available at https://lccn.loc.gov/2019006700

ISBN 13: 978-1-03-209218-8 (pbk)
ISBN 13: 978-1-138-50306-9 (hbk)

Typeset in Sabon
by Apex CoVantage, LLC

Contents

Acknowledgments

The editors, Michela Massimi and Casey D. McCoy, are grateful to all the authors who contributed to this volume for their enthusiastic engagement with the topic and the very many stimulating conversations over the past two years. Our thanks also go to Routledge editor Andrew Weckenmann for supporting this project from the beginning. This edited collection is the research output of a project that has received funding from the European Research Council (ERC) under the European Union's Horizon 2020 research and innovation program (grant agreement European Consolidator Grant H2020-ERC-2014-CoG 647272, *Perspectival Realism. Science, Knowledge, and Truth from a Human Vantage Point*). We are very grateful to the ERC for supporting our research in this area.

Illustrations

Figures

Tables

Introduction

Michela Massimi and Casey D. McCoy

Perspectivism (or perspectivalism, which one might want to call it)[1] has gained increasing attention in recent philosophy. Varieties of perspectivism have been advocated in several contexts. In epistemology, for example, Ernest Sosa (1991) originally put forward a perspectival account that was meant to go beyond the dichotomy between reliabilism and coherentism. Reflective justification for our knowledge claims is, ultimately, a matter of perspectival coherence on Sosa's view. Along similar lines, perspectival justification has been advocated by Jay Rosenberg (2002) and, in the context of the debate on peer disagreement, by Kvanvig (2013) more recently. In philosophy of language, perspectivism has featured as a tie-breaker in ongoing debates about epistemic possibilities and the limits of contextualism and relativism (see, e.g., Bach 2011). In the philosophy of time, perspectivism has found its way both in an analysis of our phenomenal experience of time (see Torrengo 2017) and in foundational analyses about the time arrow in physics (see Rovelli 2017). And when it comes to causality and the causal arrow, there too perspectivism has been presented as a promising way forward (see Price 2007; for a related discussion see Beebee 2015).

Closer to home, in philosophy of science, Ron Giere (2006) put perspectivism center stage as a viable alternative to what he portrayed as metaphysical realism and relativism in the debate about science and reality. Van Fraassen (2008) further contributed to reviving the importance of perspectival considerations in scientific representation. Ever since then, there has been a flurry of articles on the topic but no edited collection so far that directly addresses the problems and prospects of perspectivism in the philosophy of science. We hope that this book takes a first step toward remedying this lacuna. The book brings together an array of essays that reflect on perspectivism in science and shed light on the promise and limits of endorsing perspectivism in a variety of scientific fields and contexts. The ten essays here collected are all new and reflect the state of the art in this blossoming area. As maybe is to be expected in a book on perspectivism, each chapter offers a slightly different take on the topic, but the overall emphasis is on scientific challenges and methodological prospects. The former concern the viability and suitability of perspectivism to address a number of specific challenges in scientific fields such as physics, biology, cognitive neuroscience, and medicine, just as a few examples. The latter are the prospects of deploying perspectivism successfully as a methodology to tackle general issues about conceptual change and semantics, modeling and representing, the nature of measurement, pluralism, realism, and interdisciplinarity. Some of these chapters speak more directly to scientific challenges; others to the methodological prospects. Overall, this book is meant to be a kaleidoscopic journey toward a better understanding of the philosophical view called "perspectivism" and its relevance to science.

Because of its kaleidoscopic nature—and despite the lack of a common working definition of what a "scientific perspective" is across the ten essays—there are nonetheless some important common themes that run through this collection. The first is the relation between perspectivism and pragmatism. The second is the relation between perspectivism and pluralism. The third is the relation between perspectivism and realism. Let us consider each of them in turn.

Perspectivism shares with pragmatism a common origin in their respective commitment to uphold a view of scientific knowledge different from the standard (Nagelian) "view from nowhere." Their common commitment to the idea of knowledge "from a human point of view" not only sets perspectivism and pragmatism aside from more traditional views of how science tracks reality, but it also provides a common platform for a new emphasis that both perspectivism and pragmatism place on the *historicity* of our scientific knowledge—that is, on how our scientific knowledge claims formed and developed as a result of human activities and human practices over time. A closer attention paid to the historicity of our scientific knowledge goes hand in hand with a kind of *epistemic humility* that perspectivism and pragmatism equally share (no matter whether the philosophical source for it is to be found in Kant or in American pragmatism).

This epistemic humility manifests itself in the way in which perspectivism and pragmatism alike deal with the problem of what we can claim to know within the bounds of our own epistemic perspectives. In more concrete terms, how can we ever claim to know what there is, if our epistemic access is always historically situated and perspectival? How to secure reference continuity across perspectival changes, for example? Despite these important common elements, perspectivism and pragmatism differ in their commitment to truth. For where the pragmatist tradition (for lack of a better word, given the significant differences among Peirce, Dewey, and James) drastically redefined the very notion of truth to better reflect human activities (either in terms of asymptotic limit of inquiry or of operational success in human practices), perspectivism is more neutral on the topic. Perspectivism can (and typically does) share with realism a commitment to truth as correspondence with the way things are (without necessarily having to embrace the metaphysical realist's machinery about how to understand "the way things are" in terms of truthmakers, for example).

Turning to our second common theme, namely the relation between perspectivism and pluralism that several of these chapters address directly, once again it is clear that in either case there is a commitment to a plurality of points of view. Perspectivism entails pluralism: to endorse perspectivism about science entails an endorsement of a kind of pluralism about science. But not the other way around: endorsing pluralism about science does not necessarily entail endorsing perspectivism, for scientific pluralism comes in many families and flavors and not all of them would be amenable to being classified as "perspectivism" (or perspectival pluralism). Some of the methodological problems facing perspectivism concern precisely its pluralistic nature. When it comes to a plurality of scientific models in a given area or a given target system, how to resolve the tension? What is at stake in this kind of perspectival pluralism, and can partial representation help solve the tension?

But perspectivism is not just about tackling the problems that a plurality of scientific perspectives poses for scientific modeling in some areas of inquiry. Pluralism is, first and foremost, a powerful resource in the hands of perspectivists because it shows how they can accommodate and strive to integrate a plurality of explanations for the same phenomena in some areas. The challenges and prospects of interdisciplinary research often hang on the ability to integrate a plurality of scientific perspectives, each of which might only offer a partial explanation, and be nonetheless complementary with other explanations offered by other scientific perspectives. The prospects of perspectivism—qua a kind of *explanatory* pluralism, as some of these chapters suggest—should then be assessed on a case-by-case basis, depending on the specific nature of the phenomena, their explanations, and possible (or impossible) integration.

Most importantly, this primarily *epistemic* (i.e., explanatory) reading of perspectival pluralism that emerges from some of the following chapters also makes clear the nature of the ontological stance that accompanies perspectivism. Despite the temptation to read perspectivism as an ontological view about how perspectives shape "perspectival facts," the ontological pluralism that in various forms can be found in other quarters in philosophy of science is in fact absent in all the case studies here presented. Be it protein folding, cancer research, or cognitive neuroscience, what is common to all these case studies is the emphasis placed on the plurality of *epistemic goals* associated with a form of perspectival pluralism, with no risk whatsoever of sliding into a form of fact-constructivism. This is an important point to mark in this journey toward understanding perspectivism in science.

Coming to our third and final common theme, namely the relation between perspectivism and realism that is here indirectly implied in some of the major treatments of perspectivism/pragmatism and perspectivism/ pluralism, the questions that loom large are: Is perspectivism compatible with realism? Can the two be reconciled? Some of these chapters indirectly address these broader questions by looking at either semantic issues or at the role of explanation and understanding, or by situating perspectivism within a broader trend of rethinking realism in light of historicity. A trend emerges from these chapters, one that is at pains to clarify why perspectivism is not tantamount to relativism, nor is it a redundant platitude akin to a kind of instrumentalism about science. Some authors strike a middle ground for a variety of perspectival realism that is meant to spell out the nature of reference continuity across perspectival changes. Others highlight the family resemblances with structural realism and its rationale while also pointing out the relevant differences. And yet other authors consider the question as to whether perspectivism can vindicate realist commitments borne out of a suitable notion of explanation, or whether it sits more comfortably with a non-factive understanding of phenomena.

These three main common themes are intertwined in these ten chapters. They are illustrated sometimes with a wealth of details from scientific case studies and at other times with a profound sensitivity to the history of science. Ultimately it does not matter how one defines the notion of "scientific perspective" (e.g., à la Giere, or à la van Fraassen, among others; with reference to scientific models, Kuhnian paradigms, or concepts and conceptual schemes). What matters most is what perspectivism can achieve, how it enters the practice of science, the challenges it poses, and the solutions it offers.

The book opens with Chapter 1 by Hasok Chang, who lays out his version of pragmatism and uses it as a lens through which to reflect on perspectivism in its various guises. For Chang, pragmatism is a "deep or thoroughgoing empiricism," which, however, is not an independent metaphysical or epistemological doctrine but a conception of knowledge in the service of practical goal-oriented action. Such a pragmatism is allied with perspectivism, he argues, for both are rooted in an understanding of science as a humanistic activity and both see knowledge as the product of this activity. He urges a deep perspectivism on this basis, one that holds that the relation between the world and our knowledge of it is incorrigibly perspectival. A common implication of pragmatism and perspectivism, understood in this way, is the historicity of science and scientific knowledge, which accordingly motivates and grounds the integration of history and philosophy in the study of science.

Historicity and integration pervade also Melinda Bonnie Fagan's Chapter 2. Fagan situates her engagement with perspectivism in her ongoing project on interdisciplinary research. She poses the "explanatory challenge" for interdisciplinary research, which arises from the diverse explanatory commitments in different specializations being an impediment to integrated interdisciplinary explanations. Fagan notes significant parallels with ongoing debates about perspectivism in the philosophy of science. She makes use of the latter debate in taxonomizing possible relations between models and uses the parallels between the two to apply this taxonomy to the explanatory challenge.

Epistemic humility and semantic considerations for perspectival realism are the focus of Chapter 3 by Paul Teller. Teller characterizes perspectivism as the view that human knowledge is always from a particular perspective, and agrees with Giere that perspectivism gives a realist view of science. Teller argues that perspectival realism is unlike generic scientific realism, which subsumes a particular kind of semantic realism Teller calls "referential realism." Referential realism is the view that (some) terms in scientific theories have non-empty extensions about which those theories make (approximately) true statements. Referential realism, however, fails, according to Teller: the world is too complex for the extensions of terms to be determined with our meager epistemic faculties. There are simply far too many ways in which a term's extension can be determined, given our limited access to the world. Teller's alternative to this general story of reference is pragmatic in character: our terms refer directly to idealized scenarios, but we can judge successful reference to the world by assessing the extent to which our perspectival use of referential terms succeeds in ways that we care about. The upshot is that once we see that this is how reference works, it should become clear that perspectivism is the only philosophy that doesn't make the success of reference a miracle!

The prospects and challenges of reconciling perspectivism and realism are the ongoing concerns of Chapter 4 by Juha Saatsi. Saatsi advocates a modest scientific realism, which he believes can address the challenges that scientific realism faces, such as those advanced by perspectival realists. Although some of the latter urge a perspectival account of knowledge, he believes that the perspectival aspects of science are best assimilated into the realist view in what he calls "explanatory perspectives." These explanatory perspectives play an important role in something that realists should be committed to: the accumulation of scientific *understanding*. Explanatory perspectives integrate both non-factive and factive elements, including (among the former) idealizations and false metaphysical presuppositions and (among the latter) the provision of true explanations. Saatsi's explanatory perspectives are explicated with his favored account of explanation, the counterfactual-dependence account, which is grounded in real modal connections in the world. Accumulating understanding is accordingly characterized as the increasing ability to answer counterfactual what-if-things-had-been-different questions. To illustrate how his account can address some of the perspectival challenges to scientific realism, he gives an extended example, the history of physical accounts of the rainbow. Despite the various synchronic and diachronic perspectives on the rainbow, he urges that with his account we can nonetheless appreciate a steady accumulation of scientific under-standing about this phenomenon.

The relation among perspectivism, pluralism, and realism is the topic of Collin Rice's Chapter 5. Rice too is concerned to resolve the challenge presented by multiple conflicting idealized models, which afflicts accounts of the explanatory use of idealized models in science. He appeals to the notion of a "universality class," a collection of models that display similar patterns of behavior despite being heterogeneous in their physical features. The goal is to present an alternative account of how idealized models can provide scientific explanations. He argues, furthermore, that such explanations give rise to scientific understanding in a factive sense. One may yet be a realist in the face of the plurality of models in science, since explanations that rely on universality classes do in fact capture true modal information about the world.

In Chapter 6, J. E. Wolff uses models of measurements as a case for exploring two forms of scientific realism that are meant to address the problem of plurality of models in science: structural realism and perspectival realism. She distinguishes their motivations in the following way: structural realists address the plurality of models by looking for similarities, namely structural commonalities, between the models, whereas perspectival realists emphasize how differences among a plurality of models can be complementary. In comparing these realist alternatives, Wolff chooses to focus on measurement theory. She gives two reasons: (1) Measurement theory is not a "first-order" science so much as a "meta-science" concerned with the representation of measurements in any science; as such it can give a different perspective on realism than typical case studies focused on particular sciences like physics or biology. (2) Measurement theory directly treats representation as part of its remit, and representation is, of course, one of the main points of contention in the scientific realism debate. She points out how the representationalist theory of measurement can be easily seen as a form of structural realism about measurement; however, this framework depends on the prior determination that a physical attribute satisfies the right axioms specifying a particular measurement structure. It is here that perspectivalism can enter the picture, she argues, for the representational theory of measurement does not specify how the attribution of structure is to be justified. The perspectivalist can argue that the only such justification available is one from a specific scientific perspective; in this way, Wolff urges the idea that perspectival realism and structural realism are not competing realist accounts but complementary ones.

David Danks aims to face up to two threats to perspectivism in Chapter 7. First, the risk that perspectivism might collapse into an "unsafe" relativism, where knowledge claims and the like become group specific. If perspectivism is to be a view about science, Danks argues, it needs to account for the "objectivity" of science (however, that should be understood philosophically). Second, perspectivism risks appearing too "insubstantial" to be informative. Claims such as "that objects in the world are complex" or "that humans' epistemic means are very limited" are uninformative by themselves to be helpful, at least without considerable supplementation. But Danks argues that perspectivism can be safe and substantial. He identifies two sources of perspectives: concepts and goals. *Concepts* do not merely mirror the world, he notes. Rather, they actively shape or distort information received by an epistemic agent: they give rise to distinctive perspectives that depend on what our (evolving) concepts are. Danks also links perspectives with *goals*. He points out that our goals determine, to some extent, our cognitive behavior, sometimes in such a way that we ought to believe falsehoods. Thus, insofar as concepts and goals are shared, the risk of collapsing into relativism is avoided. And as they are grounded in features of human cognition, concepts and goals make perspectives substantive. Danks's version of perspectivism, notably, is not just a view of science, since concepts and goals are part of our everyday perspectivism: it is a perspectivism of everyday lives and scientific lives alike.

Integrating a plurality of perspectives is the main theme of the final three chapters. In Chapter 8, Mazviita Chirimuuta introduces a dichotomy of research programs on the motor cortex of the brain, between what she calls the "intentional perspective" and the "dynamical perspective." In questioning the relation between these perspectives, which are taken by practitioners to be in conflict, she applies Massimi's accounts of perspectival truth to see if an integrated pluralism of perspectives might be available. She argues, however, that the mutual reinforcement of knowledge claims from the different perspectives, required by Massimi's account of truth, can only be maintained at the level of each perspective's mathematical formalism and quantitative results. The standard interpretations of these perspectives lead directly to a clash between them, from which only an instrumentalist (or better, a Kantian kind of) pluralism is left possible, according to Chirimuuta. Accepting the complexity of the brain and other entities in the world, Chirimuuta favors the pluralist response to the multiple perspectives, which she argues is necessary to understand this complexity. As a realist perspectivalism is not possible in her case study, she opts for the instrumentalist route, closing with a defense of it from the charges of relativism and fictionalism.

In Chapter 9, Anya Plutynski draws on pluralist, pragmatic, and perspectival ideas in addressing the relations between different theories, models, and research traditions (her preferred unit of analysis) in cancer research. She objects to the common narrative of scientific change that sees theories and so on in competition, where one theory "vanquishes" and succeeds another. This narrative has been pushed at times in cancer research—recently, for example, in Laplane's book *Cancer Stem Cells*, which forms Plutynski's main case study. Laplane opposes her favored cancer stem cells

(CSC) theory against the "classical theory"; Plutynski, however, shows that Laplane gives a false dichotomy, for the adversarial opposition of these two theories is not based on practice-based matters of fact. Cancer is a complex, heterogeneous disease, manifesting different features on different temporal and spatial scales, a disease that one can therefore only understand from a variety of partial perspectives. The right perspective to adopt in a context depends on practitioners' purposes. This flexibility in perspectives, analogous to Stein's idea of a dialectical tension between realist and instrumentalist attitudes, best explains approaches in cancer research. Cancer theories are complementary, and not in conflict, she urges.

In Chapter 10, Sandra D. Mitchell addresses the relations between pluralities of models that arise from the partiality of representation. She defends the idea that multiple, compatible models can be integrated in a way that increases scientific knowledge. To illustrate how her "integrative pluralism" works, she shows how three perspectives on protein folding—the physical, the chemical, and the biological—can "fill out" and correct one another. Their relationship is not a one-dimensional reductive one but one of integration. On the face of it, though, different perspectival models can appear to be in conflict by using divergent methods and representations. Mitchell argues, however, that this apparent conflict is actually productive: the preservation of pluralism precisely allows for mutual integration through mutual informing and correcting.

Clearly, understanding the many faces of perspectivism in contemporary philosophy of science requires engaging both with specific problems arising in some scientific fields and more general methodological issues concerning its relations with pluralism and realism. A lot more work still needs be done to unpack the view, its potential, and prospects. We hope that this volume can provide a first important step in this direction.

Note

1. In what follows, we use the terms "perspectivism" and "perspectivalism" interchangeably, because not much hangs on using one or the other in terms of meaning. Different authors in different contexts have been using either of the two expressions interchangeably and so will we, in keeping with this practice.

References

Bach, K. 2011. "Perspectives on Possibilities: Contextualism, Relativism, or What?" In *Epistemic Modality*, edited by Egan, A., and Weatherson, B., 19–59. Oxford: Oxford University Press.

Beebee, H. 2015. "Causation, Projection, Inference and Agency." In *Passions and Projections*, edited by Johnson, R. N., and Smith, M., 25–48. Oxford: Oxford University Press.

Giere, R. N. 2006. *Scientific Perspectivism*. Chicago: University of Chicago Press.

Kvanvig, J. L. 2013. "Perspectivalism and Reflective Ascent." In *The Epistemology of Disagreement*, edited by Christensen, D., and Lackey, J., 223–242. Oxford: Oxford University Press.

Price, H. 2007. "Causal Perspectivalism." In *Causation, Physics and the Constitution of Reality*, edited by Price, H., and Corry, R., 250–292. Oxford: Clarendon Press.

Rosenberg, J. 2002. *Thinking About Knowing*. Oxford: Oxford University Press.

Rovelli, C. 2017. "Is Time's Arrow Perspectival?" In *The Philosophy of Cosmology*, edited by Chamcham, K., Silk, J., Barrow, J., and Saunders, S., 285–296. Cambridge: Cambridge University Press.

Sosa, E. 1991. *Knowledge in Perspective*. Cambridge: Cambridge University Press.

Torrengo, G. 2017. "Feeling the Passing of Time." *Journal of Philosophy* 114(4): 165–188.

van Fraassen, B. 2008. *Scientific Representation*. Oxford: Oxford University Press.

1 Pragmatism, Perspectivism, and the Historicity of Science

Hasok Chang

1 Introduction: Humanism and Science

In this chapter, I wish to shed some light on perspectivism through the lens of pragmatism, especially in relation to scientific knowledge. The initial plausibility of this exercise lies in the fact that perspectivism and pragmatism share a deep humanist impulse, which is to regard science as a thoroughly human activity, even when it is aimed at the production of the most abstract and objective kind of knowledge. (Here I am using the terms "humanist" and "humanism" much more broadly than a strict reference to Renaissance humanism would dictate.) I will begin by outlining my interpretation of pragmatism in section 2; in fact, this is the first publication in which I attempt to lay out my view of pragmatism in any detail, so some details are necessary and this will constitute the longest part of the chapter. This will be followed, in section 3, by brief reflections on the relation between pragmatism as I understand it and perspectivism in its various guises. Afterwards, in section 4, I will explore one of the most important implications of pragmatism and perspectivism, namely the *historicity* of science and scientific knowledge. A methodological advocacy of "integrated history and philosophy of science" will naturally follow.

Humanism in relation to science is a commitment to understand and promote science as something that human agents do, not as a body of knowledge that comes from accessing information about nature that exists completely apart from ourselves and our investigations. Perhaps this humanism is not such a controversial stance (its roots go at least back to Immanuel Kant), but I think there is much value in considering its meaning and implications carefully. The most important thing about humanism as I see it is not a focus on the biological species *Homo sapiens*. For enthusiasts of artificial intelligence, animal cognition, or extraterrestrial intelligence, if we find or create serious non-human intelligence worthy of an epistemology, we might even want to call such agents "human beings" too.

In the rest of this chapter I will not speak explicitly of humanism, because I want to avoid the possibility of being mistakenly seen as advocating "human chauvinism." Also, what I want to express by "humanism" can be adequately expressed by the designation of pragmatism, which I think is the best expression of humanism among existing philosophical traditions. The most fundamental point about pragmatism, as I take it, is that knowledge is created and used by intelligent beings who engage in actions in order to live better in the material and social world.

2 What Is Pragmatism?

2.1 Beyond Semantics: Pragmatism as a Philosophy of Practice

What is pragmatism, and what does it imply for the philosophy of science? It seems that pragmatism has largely fallen off the standard philosophy curriculum, so it may not be such a bad idea to start with a quick review of the standard meanings of pragmatism. Let us pick up from where today's students and general public are likely to begin. Google defines pragmatism as "an approach that evaluates theories or beliefs in terms of the success of their practical application."[1] In more and better detail, *Webster's Ninth New Collegiate Dictionary* (1986) defines pragmatism as

> an American movement in philosophy founded by C. S. Peirce and William James and marked by the doctrines that the meaning of conceptions is to be sought in their practical bearings, that the function of thought is to guide action, and that truth is preeminently to be tested by the practical consequences of belief.

This is in fact quite a good definition. The first part of it is a version of Peirce's "pragmatist maxim," paraphrased by James here:

> to attain perfect clearness in our thoughts of an object, then, we need only consider what conceivable effects of a practical kind the object may involve—what sensations we are to expect from it, and what reactions we must prepare.
>
> (James 1907, 46–7)[2]

The Peirce–James pragmatist maxim naturally led to the semantic interpretation of pragmatism, which is perhaps the dominant one today. Christopher Hookway says, "the pragmatist maxim is a distinctive rule or method for becoming reflectively clear about the contents of concepts and hypotheses: we clarify a hypothesis by identifying its practical consequences" (2016, sec. 2). In this way, pragmatism shares much with operationalism, the homegrown philosophy of the Harvard physicist Percy Bridgman, and with the verificationism that was widely taken as a core doctrine of logical positivism. This focus on meanings continues in the current pragmatist works of Robert Brandom, Huw Price, and others.

In a similar vein, James presented pragmatism as a "method for settling metaphysical disputes that otherwise might be interminable" (1907, 45). Unless some "practical difference" would follow from one or the other side's being correct, the dispute is idle. Hookway (2016, sec. 1) recalls "a memorable illustration of pragmatism in action" by James, which shows "how the pragmatist maxim enables us to defuse an apparently insoluble (albeit 'trivial') dispute." On a visit to the mountains, James's friends engaged in a "ferocious metaphysical dispute" about a squirrel that was hanging on to one side of a tree trunk while a human observer was standing on the other side. James described the dispute as follows:

> This human witness tries to get sight of the squirrel by moving rapidly round the tree, but no matter how fast he goes, the squirrel moves as fast in the opposite direction, and always keeps the tree between himself and the man, so that never a glimpse of him is caught. The resultant metaphysical problem now is this: *Does the man go round the squirrel or not?*
>
> (James 1907, 43)

James proposed to solve the problem by pointing out that which answer is correct

> depends on what you *practically mean* by "going round" the squirrel. If you mean passing from the north of him to the east, then to the south, then to the west, and then to the north of him again, obviously the man does go round him, for he occupies these successive positions. But if on the contrary you mean being first in front of him, then on the right of him, then behind him, then on his left, and finally in front again, it is quite as obvious that the man fails to go round him, for by the compensating movements the squirrel makes, he keeps his belly turned towards the man all the time, and his back turned away. Make the distinction, and there is no occasion for any farther dispute.
>
> (James 1907, 44)

In this manner, the "pragmatic method" promises to eliminate all apparently irresoluble metaphysical disputes, and rather more important ones, too.

Even though I completely endorse the semantic tradition of pragmatism, my own emphasis is different. My inclination follows Philip Kitcher's (2012, xii–xiv) warning against the "domestication" of pragmatism. Focusing on semantics is a very effective method of domestication, making pragmatism look like a rather innocuous and interesting variation on normal analytic philosophy. I want pragmatism to be a philosophy that helps us think better about how to do things, not just about what our words mean. Recall the second part of the dictionary definition of pragmatism:

"the function of thought is to guide action." Hearing the story of James's squirrel, one might wonder: "But isn't this just a matter of defining one's terms carefully? Does it really have anything to do with pragmatism?" My take on that question is that the disambiguation offered by James is tied closely to potential practical ends. If my objective is to make a fence to enclose the squirrel, then I have gone around the squirrel *in the relevant sense*; if the objective is to check whether the wound on his back has healed, then I have failed to go around the squirrel in the relevant sense. It is the pragmatic purpose that tells us which meaning of "going round" we *ought* to mean.

2.2 Pragmatism as Empiricist Realism

One very important reason why people often do not like to go beyond the semantic dimension of pragmatism is the fear of what happens if we go further and adopt the pragmatist theory of truth. This issue needs to be tackled head-on. It is a core part of my interpretation of pragmatism that we should reject the common misperception and prejudice that pragmatism just means taking whatever is convenient as true. The "pragmatic theory of truth" attributed to James is widely regarded as absurd, and this has contributed greatly to the disdain for pragmatism among tough-minded philosophers. Here is probably the most notorious statement by James: "*'The true,' to put it very briefly, is only the expedient in the way of our thinking, just as 'the right' is only the expedient in the way of our behaving*. Expedient in almost any fashion" (James 1907, 222). I think James's choice of the word "expedient" was unfortunate, as sounding too much like just "convenient" or "useful"—or perhaps the word had quite a different connotation back then; that is for James scholars to debate. At any rate, the statement actually continues as follows:

> And expedient in the long run and on the whole of course; for what meets expediently all the experience in sight won't necessarily meet all farther experiences equally satisfactorily. Experience, as we know, has ways of *boiling over*, and making us correct our present formulas.
> (James 1907, 222)

I want to argue that what this passage really shows is James the staunch empiricist, declaring that the source of truth is experience, and that it is futile to entertain any more grandiose notion of truth. This provides an important clue to my interpretation of pragmatism. My proposal is to understand pragmatism as a deep or thoroughgoing empiricism, which recognizes experience as the only ultimate source of learning and refuses to acknowledge any higher authority. Something does need to be said in justification of empiricism, but for now let me take it as a credo, as an article of faith; some sort of empiricism might be the inevitable starting point of epistemology in our scientific age, as much as the presumption of God would have been the inevitable bedrock of any intellectual discourse in Europe in an earlier age.

The spirit of empiricism has been summarized rather poetically by Clarence Irving Lewis, in his review of John Dewey's 1929 masterpiece, *The Quest for Certainty*:

> Man may not reach the goal of his quest for security by any flight to another world—neither to that other world of the religious mystic, nor to that realm of transcendent ideas and eternal values which is its philosophical counterpart. Salvation is through work; through experimental effort, intelligently directed to an actual human future.
> (Lewis 1930, 14)

This passage is especially nice because it brings together the two pragmatist philosophers that I have found most inspiring.

On such an empiricist conception of knowledge, how might we make sense of traditional epistemic and metaphysical notions such as truth and reality? Central to my thinking is the notion of *operational coherence*, a harmonious fitting-together of actions that is conducive to a successful achievement of one's aims.[3] To put it somewhat more precisely: an activity is operationally coherent if and only if there is a harmonious relationship among the operations that constitute the activity. The concrete realization of a coherent activity is successful *ceteris paribus*; this serves as an indirect criterion for the judgment of coherence. Operational coherence pertains to an epistemic activity (or a system of practice), not to a set of propositions; it is measured against the aims of the activity (or system) in question. Coherence may be exhibited in something as simple as the correct coordination of bodily movements needed in lighting a match or walking up the stairs, or something as complex as the successful integration of a range of material technologies and various abstract theories in the operation of the Global Positioning System. It has social and emotional aspects as well as material and intellectual ones.

Coherence is the chief characteristic underlying a successful epistemic activity. It is the vehicle through which the mind-independent world is brought to bear on our knowledge. Operational coherence carries within it the constraint by nature, and in fact it is the only way in which reality can give input to our knowledge. Using this notion of coherence, I propose a new coherence theory of truth: a statement is true in a given circumstance if (belief in) it is needed in a coherent activity (or system of practice).[4] Truth understood in this way comes with a specific scope or domain attached to it in each case, which allows us to legitimize intuitive statements such as "Newtonian mechanics remains true in the domain of macroscopic objects moving at low velocities." And because coherence is a matter of degree, so is truth—and I think that is also right. J. L. Austin noted long ago (1979, 117, 130–131) that "very true," "true enough," and such are perfectly sensible locutions. Catherine Elgin (2017) has more recently shown the pragmatic power of "true enough" accounts. It is not necessary to conceive of truth itself as a binary yes-no property,[5] and insist on speaking in terms of approximate truth or partial truth when we wish to discuss degrees of truth. The notion of (empirical) truth I propose can ground a kind of realism that is not at all contrary to empiricism.

2.3 The Empirical Learning of Methods

One salient feature of the deep empiricism that I see in pragmatism is empiricism concerning methods, which received its full articulation in Dewey's late work *Logic*, which he strikingly subtitled *The Theory of Inquiry*. According to Dewey (1938, 12), scientific methods and logical rules arise from successful habits of thinking. Content and method are learned through the same process of inquiry. Success is being "operative in a manner that tends in the long run, or in the continuity of inquiry, to yield results that are either confirmed in further inquiry or that are corrected by use of the same procedures" (13). This is how method-learning happens:

> through comparison-contrast, we ascertain *how* and *why* certain means and agencies have provided warrantably assertible conclusions, while others have not and *cannot* do so in the sense in which "cannot" expresses an intrinsic incompatibility between means used and consequences attained.
>
> (104)

And "we know that some methods of inquiry are better than others in just the same way in which we know that some methods of surgery, farming, road-making, navigating or what-not are better than others" (104).

Dewey stressed the continuity of rules—of logic, everyday inquiry, and the scientific method (4–6). According to him, even logical rules, like any other rules, receive their justification through the success of inquiry. He considered that "all logical forms (with their characteristic properties) arise within the operation of inquiry, and are concerned with the control of inquiry so that it may yield warranted assertions" (4).[6] What he called the "fundamental thesis" of his book was that "logical forms accrue to subject-matter when the latter is subjected to controlled inquiry."[7] So it was crucial to look at the history of logic, which he regarded as a progressive discipline. Dewey argued that we needed logic to suit the modern scientific way of thinking, and lamented that logicians of his day tended to retain the form of classical logic while abandoning the metaphysical and operational underpinnings of it.[8] In chapter 5 of *Logic*, he undertook a contextual analysis of Aristotelian logic, arguing that it was a system admirably suited for the science and philosophy of ancient Greece (perhaps only Athens), but no longer suited to the mode of scientific and philosophical thinking, which had changed dramatically since then. As some key elements of Aristotelian thinking that had been abandoned, he identified essentialism, the emphasis on quality over quantity, static classification as the form of knowledge, and the heterogeneous and hierarchical structure of the universe.

2.4 The Empirical Validation of the A Priori

What I am calling the deep empiricism of the pragmatists even touched upon the nature of a priori knowledge, as already indicated by Dewey's views on logic. On this point, the most eloquent exponent of pragmatism was Lewis who, according to L. W. Beck, once declared "I am a Kantian who disagrees with every sentence of the *Critique of Pure Reason*" (in Schilpp 1968, 273). The core of Lewis's disagreement with Kant was that Lewis denied the existence of synthetic a priori judgments. He did think that there was an important a priori element to knowledge, but that it was always analytic: "*The a priori is not a material truth, delimiting or delineating the content of experience as such, but is definitive or analytic in its nature*" (Lewis 1929, 231). A priori propositions are inherent in "conceptual systems," but these systems are constructed and chosen by us on "instrumental or pragmatic" grounds (x). Once we have chosen a conceptual system, within it the a priori elements are analytically true.[9]

Lewis called his position "conceptual[istic] pragmatism" and acknowledged his debt to James, Dewey, and especially Peirce, but signaled a distance from "orthodox" pragmatism (xi). However, I think Lewis's take on the a priori deserves to be brought into the mainstream of pragmatism. It is the epitome of deep empiricism that even the adoption of the a priori is made on empirical grounds. Lewis explains:

> While the *a priori* is *dictated* neither by what is presented in experience nor by any transcendent and eternal factor of human nature, it still *answers* to criteria of the general type which may be termed pragmatic. The human animal with his needs and interests confronts an experience in which these must be satisfied, if at all. Both the general character of the experience and the nature of the animal will be reflected in the mode of behavior which marks this attempt to realize his ends. This will be true of the categories of his thinking as in other things. And here, as elsewhere, the result will be reached by a process in which attitudes tentatively assumed, disappointment in the ends to be realized, and consequent alteration of behavior will play their part.
>
> (239, emphases added)

This pragmatic nature of the a priori also makes it historical, as various neo-Kantian thinkers ranging from William Whewell to Michael Friedman have emphasized: "there will be no assurance that what is a priori will remain fixed and absolute throughout the history of the race or for the developing individual"; "if the a priori is something made by the mind, mind may also alter it"; "the determination of the a priori is in some sense like free choice and deliberate action" (233–234). In this connection, Lewis pays some attention to the actual history of science:

> New ranges of experience such as those due to the invention of the telescope and microscope have actually led to alteration of our categories in historic time. The same thing may happen through more penetrating or adequate analysis of old types of experience—witness Virchow's redefinition of disease. What was previously regarded as real—e.g., disease entities—may come to be looked upon as unreal, and what was previously taken to be unreal—e.g., curved space—may be admitted to reality. But when this happens *the truth remains unaltered and new truth and old truth do not contradict*. Categories and concepts do not literally change; they are simply given up and replaced by new ones.
>
> (268)

It is interesting to consider how Lewis's view on the a priori extends to logic. Lewis (vii) himself said that his pragmatist epistemology in fact arose from his work in symbolic logic, which was highly respected. There *are* different systems of logic, and anyone who wants to reason logically must start by adopting a particular system of logic. But the only plausible and non-arbitrary way of justifying the choice of a logical system would be on pragmatic grounds, because appealing to the rules of logic for this choice would clearly be question-begging. So it may actually turn out that the treatment of logic is the most convincing part of pragmatism! With Lewis's contribution, and the current proliferation of non-classical logics and their successful application in the design of intelligent systems, I think we must admit that Dewey has been vindicated in his fundamental thesis: "Logical forms accrue to subject-matter when the latter is subjected to controlled inquiry" (101). And then it was perhaps natural for Lewis to generalize such thoughts about logic to say that the justification of the choice of any conceptual system can only accrue from the experience of trying to apply the system in question to various areas of inquiry (x–xi).

2.5 The Active Nature of Knowledge

Having considered what pragmatism should mean, we can attempt an overview of the fundamental question of epistemology: what is the nature of knowledge? Pragmatism as I see it does not sit well with the

propositional conception of knowledge that forms the bedrock of epistemology in the analytic tradition. With its emphasis on method-learning
and practices of inquiry, pragmatism needs a conception of knowledge
understood as an ability, the ability to achieve certain aims. The propositional conception of knowledge was quite explicitly criticized by James
and Lewis when they attacked the "copy theory" of knowledge. According to James:

> The popular notion is that a true idea must copy its reality. Like other
> popular views, this one follows the analogy of the most usual expe-
> rience. Our true ideas of sensible things do indeed copy them. Shut
> your eyes and think of yonder clock on the wall, and you get just
> such a true picture or copy of its dial. But your idea of its "works"
> (unless you are a clockmaker) is much less of a copy, yet it passes
> muster, for it in no way clashes with reality. Even though it should
> shrink to the mere word "works," that word still serves you truly;
> and when you speak of the "time-keeping function" of the clock, or
> of its spring's "elasticity," it is hard to see exactly what your ideas
> can copy.
>
> (1907, 199)

Lewis was more succinct: "Knowledge does not copy anything presented.
It proceeds from something given toward something else. When it finds
that something else, the perception is verified" (1929, 162).[10] Here Lewis
is echoing Dewey's notion of inquiry, viewed thoroughly and explicitly
as a process: "*inquiry is the controlled or directed transformation of an
indeterminate situation into one that is so determinate in its constituent
distinctions and relations as to convert the elements of the original situation into a unified whole*" (Dewey 1938, 104–105).[11] Here it may be
argued that Dewey was developing a notion from Peirce, who in "The
Fixation of Belief" had set out a perspective on inquiry as a process in
which we "struggle to free ourselves" from a state of doubt and "pass
into the state of belief" (1877, 5). Even though Peirce tended not to
focus on the practices of inquiry, when he did comment on them he was
clear about their processual dimension. And Cheryl Misak adds that the
important thing Peirce regarded as wrong with the state of doubt is "that
it leads to a paralysis of action" (2013, 33).

"The knower is an actor," declared James (quoted in Putnam 1995, 17).
Dewey went on to develop this vision fully, complete with his own
memorable slogan: "we live forward" (Dewey 1917, 12). Experience is
active, full of expectations and reactions, contrary to the impoverished
view of it in traditional empiricism as the recording of information.
Experience, and knowledge too, is something taken by active agents.
Inquiry is pervasive in life, an essential activity of an organism coping
in its environment.

A pragmatist philosophy of science should recognize clearly that scientific inquiry is itself a kind of human experience. Learning from experience also requires learning about the nature of that experience of learning. Pragmatist philosophical attention needs to be turned to the process of knowledge-production itself in order to provide an elucidation of epistemic activities. What do we do in order to gain knowledge, to test it, to improve it, to use it? How best do we organize and support such epistemic acts that we engage in? If we conceive of pragmatism generally as a philosophical commitment to engage with practices, then pragmatist epistemology will concern itself with all practices relating to knowledge. I believe that this is something that pragmatists were seriously engaged in.

For my own thinking about scientific practice, I have devised notions of "epistemic activities" and "systems of practice" (Chang 2014). What *kinds* of things do scientists do when they do science? They engage in some very complex practices, which may be analyzed as composites of many different kinds of basic epistemic activities, such as describing, predicting, explaining, hypothesizing, testing, observing, detecting, measuring, classifying, representing, modeling, simulating, synthesizing, analyzing, abstracting, and idealizing. Many of these categories may seem like simple types of mental acts, but when we ask how any of it is actually done in particular situations, we discover that a complex network of material, mental, and social activities are involved. As an illustration, take something that would seem very far removed from actions: the definition of a concept. Consider what one has to *do* in order to define a scientific term: formulate formal conditions for its correct verbal and mathematical use; construct physical instruments and procedures for measurement, standard tests, and other manipulations; round people up on a committee to monitor the agreed uses of the concept and devise methods to give penalties to people who do not adhere to the agreed uses. In one stroke, we have brought into consideration all kinds of unexpected things, ranging from operationalism to the sociology of scientific institutions. "One meter" or "one kilogram" would not and could not mean what it means without a whole variety of epistemic actions coordinated by the International Bureau of Weights and Measures in Paris. Even semantics is a matter of doing, as Wittgenstein and Bridgman taught us long ago.

3 Pragmatism and Perspectivism

Having sketched my own interpretation of pragmatism, I now want to consider how it relates to perspectivism. As indicated at the start by reference to humanism, perspectivism has a great deal of affinity with pragmatism as I see it. They are both rooted in taking science as an activity carried out by humans or other intelligent *agents*, and scientific knowledge as a product of such an activity. Any knowledge arising from a particular activity will bear a clear contextual or perspectival stamp of its origins. The only real difference may be in emphasis, pragmatism being more explicit than perspectivism in its action-orientation.

But what exactly is perspectivism? I start by following the recent exposition by Michela Massimi (2018), in which she consciously builds on the earlier articulation by Ronald Giere (2006). Overall, she defines perspectivism (or perspectival realism) in the philosophy of science as "a family of positions that in different ways place emphasis *on our scientific knowledge being situated*." There are two main ways of being situated:

(1) Our scientific knowledge is *historically situated*, that is, it is the inevitable product of the historical period to which those scientific representations, modeling practices, data gathering, and scientific theories belong.
And/Or
(2) Our scientific knowledge is *culturally situated*, that is, it is the inevitable product of the prevailing cultural tradition in which those scientific representations, modeling practices, data gathering, and scientific theories were formulated.

(Massimi 2018, 164)

As Massimi's definition indicates, there is no consensus on the precise interpretation of perspectivism. And what I want to do, especially in the light of various considerations made above in my interpretation of pragmatism, is to advocate a rather strong form of perspectivism—Kitcher's warning against domestication should be applied to perspectivism as well as pragmatism. Massimi goes part of the way with me here in stating that "truth-conditions for scientific knowledge claims vary in interesting ways depending on the context in which they are uttered and used" (171). I could not agree more, but I do not think it goes far enough. It is not only the truth conditions for a knowledge claim that are perspectival but the knowledge claims themselves. Even if we just take the semantic version of pragmatism, the very meanings of any concepts or terms we use are only contextually fixed. So there cannot be any knowledge claim that is not perspectival. Now, I may agree with the common notion endorsed by Massimi that "there are perspective-independent worldly states of affairs that ultimately make our scientific knowledge claims true or false"; however, I remain with Kant in insisting that such states of affairs, as such, are not expressible.[12] So it is not only "*our ability to know* these states of affairs" that "depends inevitably on the perspectival circumstances or context of use" (171), but—I further argue—the very possibility of articulating anything about the state of the world. All we can ever talk about are conceptualized objects, which are in the realm of Kantian phenomena rather than things-in-themselves.

It might be useful to lay out here the three separate layers of perspectivism that I see. This is my own perspective, as it were, on perspectivism.

Each of the three layers is compatible with some version of pragmatism. However, my own reading of pragmatism tends to go with the third and deepest layer of perspectivism.

(1) The same content can be expressed in different ways—in different languages, or using different expressions, that are not incommensurable with each other. In such cases, it can be considered that the strict meaning of the different expressions is the same. However, there will typically be different connotations attached to the different expressions, and such differences in connotation can also embody divergent expectations, and can prompt and facilitate divergent courses of action. For example, consider how classical mechanics can be expressed in the Newtonian, Lagrangian, and Hamiltonian formulations. Even though it can be shown that these formulations are formally equivalent to each other, there are very significant practical consequences in problem-solving and further theorizing that follow from the choice.

(2) Different perspectives can make us highlight and focus on different aspects of a given object, and can also blind us to other aspects. This sense of perspectivism is consonant with quite a literal reading of "perspective": if we look at a three-dimensional object in the normal human way, we will only see a two-dimensional picture whose content depends on the direction from which we are looking. Or we can project very different-looking shadows of one and the same three-dimensional object in different directions. A similar image of "viewing objects or scenes from different places" is the device with which Giere (2006, 13) initially introduces the idea of perspectivism in his book. Generalizing this thought in a metaphorical way to the role that conceptual frameworks play in our cognition, we say that we can learn about something in different perspectival ways, like the proverbial blind people feeling different parts of an elephant. On a large scale, Giere (2006, chap. 4) takes it that scientific theories or models provide such perspectival representations of the world as maps based on different projection methods do with the earth. This level of perspectivism still allows the possibility that we can build a true picture of the object, by unifying enough well-placed perspectival pictures of it, as we can similarly construct a three-dimensional image of an organ in a CT scan based on various two-dimensional cross-sections taken with x-rays. This view is perfectly compatible with the standard sort of scientific realism.

(3) Going more deeply perspectival, one can argue that the relation between our knowledge and the world cannot be spelled out in a straightforward way as given in the two above interpretations of perspectivism. Projection is a metaphor, as is "perspective" itself. The very seductive and deeply misleading aspect of those metaphors is that we take it for granted that the three-dimensional objects we are perspectively studying exist "out there" in themselves, well-formed independently of all our cognition and action. When we are facing the universe, we do not

have the equivalent of the perfectly understood three-dimensional object that we try to represent from various two-dimensional perspectives. The strong pragmatism I have articulated argues against the cogency of such a notion of ready-made reality and affirms the strong ontological perspectivism that Anjan Chakravartty warns us against: "there are no perspective-transcendent ontological facts or states of affairs" (2017, 177). People often imagine that the broadly accepted "scientific" picture of the world is the ready-made reality, rather than just one perspectival picture; this is to commit the error of "hypostatization," as Dewey named it.[13] Any phenomenon of nature that we can think or talk about at all is couched in concepts, and we choose from different conceptual frameworks (as C. I. Lewis emphasized), which are liable to be incommensurable with each other. If we take "perspective" to mean a conceptual framework in this sense, then we can see that ontology itself is perspectival. Only unspeakable noumena may be imagined to exist in an absolute sense, guaranteed to be shared between incommensurable frameworks.

The kind of realism sanctioned by the last kind of perspectivism is strongly pluralist: each good perspective offers a true account that is worth preserving and developing, without the need or sometimes even the possibility to reduce or bind it to another perspective (Chang 2018). The knowledge gained from each perspective answers to reality in its own way, but not in a way that is straightforwardly comparable to other ways. All this is not to deny that it may be possible to make productive links, sometimes even reductive or unificatory ones, between different perspectives; however, whether such links are possible is an empirical question—neither a foregone conclusion nor an inescapable imperative. The pluralism expressed here can resist the usual attack on perspectivism through the charge of relativism. A recent, well-reasoned instance of such an attack is Chakravartty's book on "scientific ontology," where he denounces the "Kant on wheels" type of ontology as incoherent (2017, 186). Yet Chakravartty actually seems to be a strong ontological pragmatist and pluralist:

> We know only about those ontological packages that have proven sufficiently successful as posits in these particular contexts . . . nothing in scientific practice precludes the existence of other packages and behaviors that are unknown to us and, indeed, the hubristic image of our own ontological powers that would be required to think otherwise should arguably give one pause.
>
> (196–197)

It is in my view immaterial whether or not he likes to use the label of "perspectivism" to characterize this view.

To sum up: pragmatism, as I articulate it in this chapter, strengthens perspectivism. Not only is such pragmatism consonant with various versions of perspectivism (partly due to the humanist grounding that they both share), but it should give perspectivists the courage to see just how deeply perspectivism can and should go.

4 The Historicity of Science and the Need for Integrated HPS

If we take seriously the pragmatist or perspectivist view of knowledge and inquiry as articulated here, what follows about the nature of science? And what does that imply about how we should practice the history and philosophy of science (HPS)? Here I return to the idea of pragmatism as deep empiricism. If the only learning we can do comes from experience, then learning about how to learn can only come from a study of the history of learning. If philosophy of science concerns itself with understanding and improving the methods of science, then it can only function properly by following the history of science. So, the epistemological side of philosophy of science is inseparably tied to history of science. This is an intuition that most practitioners of HPS already have, but I believe that the intuition should be grounded in a clearer and more explicit conception of knowledge.

It is an empirical fact that humans have learned that there are some valid patterns of inquiry that are quite common in various kinds of situations. As Dewey put it, "inquiry, in spite of the diverse subjects to which it applies . . . has a common structure or pattern . . . applied both in common sense and science" (1938, 101). In Dewey's view, there is a continuity among various types of rules that govern everyday inquiry, scientific practice, and logical inference (4–6); scientific methods and logical rules arise from successful habits of thinking, and they receive their justification through the success of inquiry (12).

Another observation to add to the basic historicity of learning is that truly empirical processes of learning need to be iterative. I had acquired a dim awareness of this from my work on the development of temperature measurement standards, through which I came up with the notion of "epistemic iteration" with a little bit of inspiration from Peirce (Chang 2004, 44–48, 220–234). But Dewey had seen the iterative nature of learning in a very general light. The key question, according to Dewey, was "whether inquiry can develop in its own ongoing course the logical standards and forms to which *further* inquiry shall submit." And "one might reply by saying that it can because it has" (Dewey 1938, 5). Because there are no "standards *ab extra*" that inquiry can rely on, it must be possible to have a "self-corrective process of inquiry" if we are to get anywhere. Logical principles, at each stage of this process, are "operationally *a priori*" (14). This view is very consonant with that of Lewis about a priori principles in general.

Now consider the nature of knowledge as well as inquiry. If knowledge is the product of inquiry, and inquiry is an iterative process as we have just described, then knowledge can only be understood fully by knowing the history of inquiry, because any stage of knowledge can only develop by relying on the previous stage, and its shape is ineliminably influenced by its starting point. It is not good enough to know the current conditions of science, because present science is not only a response to present needs, but it is in large part also determined by our inheritance, which arose from answering past needs.

Learning about the past conceptions and techniques that have shaped our present science is not just a matter of description. It is actually much more crucial for a normative project. In order to re-process our inheritance intelligently, we should first understand how it came into being. Benedetto Croce once said, "only historical judgment liberates the spirit from the pressure of the past" (1941, 48). Learning the history of how scientists came to believe today's orthodoxy can not only aid our understanding of the science, but it may also reveal that what we take as evident and necessary truths today were results of past decisions that were contingent and could have gone in a different way. Such awareness of past contingency also allows the possibility of making that contingency real again in the present; thereby history of science can legitimately support normative philosophy of science too, quite contrary to common impressions of the fundamentally descriptive nature and mission of historiography.

In the converse direction of dependence, why the history of science needs the philosophy of science will perhaps be quite obvious to most perspectivist readers of this volume. No history can be written as a "view from nowhere." And no history of science can be written in a philosophy-of-science vacuum, if we take philosophy of science as a field that provides notions about the nature of scientific practice and scientific knowledge. The question is what kind of philosophy we should take as our historiographical framing device. The old internalism, which was certainly amenable to history–philosophy integration, achieved the integration only in a diminished fashion. It tended to talk of knowledge as theories and experiments existing in the service of theories. It was linked up with a truth-focused justificationist epistemology firmly based on the propositional conception of knowledge. This kind of history–philosophy integration was only made possible by impoverishing the history. Pragmatism and perspectivism offer superior alternatives.

Here I return to the broad humanist framing with which I began this chapter. If we appreciate the acquisition and development of knowledge fully in the context of human life, we need to examine the development of scientific knowledge that serves the purposes of life. Recall the humanist flourish of the Vienna Circle Manifesto:

Everything is accessible to man; and man is the measure of all things. Here is an affinity with the Sophists, not with the Platonists; with the Epicureans, not with the Pythagoreans; with all those who stand for earthly being and the here and now.

(Neurath et al. 1973, 306)

"The scientific world-conception serves life, and life receives it" (318). Such an outlook on science can best be framed not in terms of logical positivism as it is commonly understood, but in terms of pragmatism taken as deep empiricism.

There are many reasons to pursue integrated HPS, and many ways for doing it.[14] What has been lacking in my own advocacy of integrated HPS is a broad philosophical grounding for it, a general philosophical perspective from which integrated HPS is a crucial thing to be doing for those who concern themselves with the place of science in human life. The need for integrated HPS arises naturally if we take a humanist view of science.

Acknowledgments

This chapter originated as a keynote address at the Sixth International Conference on Integrated History and Philosophy of Science ("&HPS6") in Edinburgh under the title of "Pragmatism, Humanism and Integrated HPS." I thank Michela Massimi for inviting me to give that presentation and to adapt it as a chapter in the present volume. Our collaboration pertinent to the generation of this chapter dates back to my participation in her research program on Kant and the philosophy of science, undertaken while we were colleagues in the Department of Science and Technology Studies at University College London (UCL). I also thank other colleagues in the Committee for Integrated HPS. My ideas about pragmatism have been developed through presentations and seminars in Bielefeld, Tartu, Vienna, Cambridge, UCL, and Hanyang University; I thank all my hosts and audiences there, especially Ave Mets and Martin Carrier.

Notes

1. This definition appears on simply searching for "pragmatism" in Google, and it is part of the Google Dictionary (www.google.be/search?q=Dictionary).
2. For the original formulation, see Peirce (1878, 293).
3. For some preliminary thoughts along this direction published so far, see Chang (2016, 2017, 2018). In Chang (2016), I used the phrase "pragmatic coherence."
4. A preliminary exposition of this account is in Chang (2017). The epistemic activity in question does not have to be one of explicit testing. Sometimes a true statement is explicitly verified; other times its truth consists in its involvement in other kinds of successful activities.
5. Many-valued logics are possible, and in fact seem to be flourishing.

6. Dewey references Peirce in connection with this thought, in footnote 1 on page 9.
7. That was the first of his "six theses" about logic (Dewey 1938, 14–21).
8. Dewey argued that logical principles were only "operationally a priori" at each stage of the evolution of knowledge (Dewey 1938, 14). If that is the situation even with logic, need we say much about the revisability and evolvability of the methods of science?
9. The following statements are helpful in clarifying Lewis's position: "The necessity of the a priori is its character as legislative act. It represents a constraint imposed by the mind, not a constraint imposed upon mind by something else" (Lewis 1929, 197). "The paradigm of the *a priori* in general is the definition. It has always been clear that the simplest and most obvious case of truth which can be known in advance of experience is the explicative proposition and those consequences of definition which can be derived by purely logical analysis. These are necessarily true, true under all possible circumstances, because definition is legislative" (239–240).
10. See also Lewis (1929, 165).
11. Dewey went on to dissect the inquiry process into several steps (Dewey 1938, 105–112).
12. There are, of course, intricate debates about how much Kant was willing to enter into metaphysical commitments reaching beyond the phenomenal realm. See, for example, Massimi (2017) for discussion.
13. See Fesmire (2015, 64–73) for a nice exposition.
14. See Arabatzis and Schickore (2012), and references therein, for a relatively recent assessment of the state of play in this area of work.

References

Arabatzis, T., and Schickore, J. 2012. "Ways of Integrating History and Philosophy of Science." *Perspectives on Science* 20(4): 395–408.

Austin, J. L. 1979. "Truth." In *Philosophical Papers*, 3rd ed., edited by Urmson, J. O., and Warnock, J. G., 117–133. Oxford: Oxford University Press.

Chakravartty, A. 2017. *Scientific Ontology*. New York: Oxford University Press.

Chang, H. 2004. *Inventing Temperature: Measurement and Scientific Progress*. New York: Oxford University Press.

Chang, H. 2014. "Epistemic Activities and Systems of Practice: Units of Analysis in Philosophy of Science After the Practice Turn." In *Science After the Practice Turn in the Philosophy, History and Social Studies of Science*, edited by Soler, L., Zwart, S., Lynch, M., and Israel-Jost, V., 67–79. London and Abingdon: Routledge.

Chang, H. 2016. "Pragmatic Realism." *Revista de Humanidades de Valparaíso* 4(2): 107–122.

Chang, H. 2017. "Operational Coherence as the Source of Truth." *Proceedings of the Aristotelian Society* 117(2): 103–122.

Chang, H. 2018. "Is Pluralism Compatible With Scientific Realism?" In *The Routledge Handbook of Scientific Realism*, edited by Saatsi, J., 176–186. London and New York: Routledge.

Croce, B. 1941. *History as the Story of Liberty*, translated by Sprigge, S. New York: Norton.

Dewey, J. 1917. "The Need for a Recovery of Philosophy." In *Creative Intelligence: Essays in the Pragmatic Attitude*, edited by Dewey, J., 3–69. New York: Henry Holt and Company.

Dewey, J. 1938. *Logic: The Theory of Inquiry*. New York: Henry Holt and Company.

Elgin, C. Z. 2017. *True Enough*. Cambridge, MA: MIT Press.

Fesmire, S. 2015. *Dewey*. London and New York: Routledge.

Giere, R. 2006. *Scientific Perspectivism*. Chicago: University of Chicago Press.

Hookway, C. 2016. "Pragmatism." In *The Stanford Encyclopedia of Philosophy*, Summer 2016 ed., edited by Zalta, E. N. https://plato.stanford.edu/archives/sum2016/entries/pragmatism/.

James, W. 1907. *Pragmatism: A New Name for Some Old Ways of Thinking*. New York: Longmans, Green, and Company.

Kitcher, P. 2012. *Preludes to Pragmatism: Toward a Reconstruction of Philosophy*. New York: Oxford University Press.

Lewis, C. I. 1929. *Mind and the World-Order: Outline of a Theory of Knowledge*. New York: Charles Scribner's Sons.

Lewis, C. I. 1930. "The Quest for Certainty: A Study of the Relation of Knowledge and Action by John Dewey." *Journal of Philosophy* 27(1): 4–25.

Massimi, M. 2017. "Grounds, Modality, and Nomic Necessity in the Critical Kant." In *Kant and the Laws of Nature*, edited by Massimi, M., and Breitenbach, A., 150–170. Cambridge: Cambridge University Press.

Massimi, M. 2018. "Perspectivism." In *The Routledge Handbook of Scientific Realism*, edited by Saatsi, J., 164–175. London and New York: Routledge.

Misak, C. 2013. *The American Pragmatists*. Oxford: Oxford University Press.

Neurath, O., et al. 1973. "Wissenschaftliche Weltauffassung: Der Wiener Kreis." In *Empiricism and Sociology*, edited by Neurath, M., and Cohen, R. S., 299–318. Dordrecht: Springer.

Peirce, C. S. 1877. "The Fixation of Belief." *Popular Science Monthly* 12(November 1877): 1–15.

Peirce, C. S. 1878. "How to Make Our Ideas Clear." *Popular Science Monthly* 12(January 1878): 286–302.

Putnam, H. 1995. *Pragmatism: An Open Question*. Oxford: Blackwell.

Schilpp, P. A., ed. 1968. *The Philosophy of C. I. Lewis*. La Salle: Open Court.

2 Explanation, Interdisciplinarity, and Perspectives

Melinda Bonnie Fagan

1 Introduction

This chapter engages with perspectivism as part of an ongoing project on explanation as collaboration. The latter project requires some introduction, which will set the stage for my engagement with perspectivist ideas. The goal of this chapter, hence, is twofold. First, I show that a focal problem for my project, the explanatory challenge for interdisciplinary research, has significant parallels with recent debates about perspectivism. Second, I use those parallels to construct a taxonomy of possible relations between models in interdisciplinary research. The result is a repurposing of the perspectivism debate as a step toward answering the explanatory challenge for interdisciplinary research. I conclude with a sketch of the next steps in the project.

The remainder of this section provides the necessary background. The project on explanation as collaboration is a contribution to philosophy of science in practice, an increasingly prominent approach that examines the actual practices of scientists rather than philosophical idealizations or reconstructions. The starting point for this project is a broad view of scientific practice as socially complex, comprising many diverse and dynamically interacting lines of inquiry. Three features of science in practice are of particular interest for my purposes:

1. Scientific practice is very heterogeneous, both over time and across specializations.
2. Scientific specializations differ from one another in their epistemic goals, methods, norms, and products.
3. Among the most distinctive products of a scientific specialization are its explanations.

These three features set the stage for the explanatory challenge for interdisciplinary research, which is presented in the next section. Here, I provide a bit more background in order to clearly indicate the stance from which I approach scientific perspectivism.

At any given time, scientific practice as a whole consists of a multitude of diverse specializations, many proceeding more or less independently from others, with some in the process of merging or splitting. With increasing numbers of researchers and more elaborate technology and training programs, the rate of specialization splitting is increasing.[1] What I term a specialization is very similar to Darden and Maull's classic account of a scientific field:

> a field is an area of science consisting of the following elements: a central problem, a domain consisting of items taken to be facts relating to that problem, general explanatory factors and goals providing expectations as to how the problem is to be solved, techniques and methods, and sometimes, but not always, concepts, laws and theories which are related to the problem and which attempt to realize the explanatory goals.
>
> (Darden and Maull 1977, 44)

There are two differences between Darden and Maull's notion of a field and a specialization, as I use that term here. First, their concept of a field is conceptual and historical, not sociological. A specialization, as I conceive it, is a sociological entity as well as a conceptual and historical one, comprising multiple epistemic agents interacting with one another in a cultural context. Second, Darden and Maull define a field as aimed at explanation, conceived as a form of problem-solving. I relax the assumption that explanation is a necessary scientific goal, allowing that some specializations may not aim at explanation. Not all problem solutions in science involve explanation, though many do. I retain, however, Darden and Maull's focus on explanation—not because explanation is all there is to scientific practice, but because it is an important part of that practice and, moreover, one with social (collaborative) dimensions that have not been sufficiently appreciated. Explicating these social dimensions is the overall goal of my larger project on explanation and collaboration.

An explanation, as I use the term, is a product of scientific inquiry, constructed with a set of tools and in accordance with norms of a specialization, on which the honorific "explanation" is bestowed by members of that specialization. On this view an explanation is an epistemic artifact, produced and recognized by a scientific community. This approach allows for many different styles and forms of explanation; it is, so to speak, an actor's category for scientific communities. This starting point excludes ontic accounts of explanation that locate explanatory factors in the world itself rather than in our representations of it. That exclusion is not a problem for my purposes, however. This is because any explanation provided by the world must be interpreted by scientists in order to play a role in scientific practice. Such an interpretation, including reference to the world independent of human thought and action, if endorsed by an actual scientific specialization, is an explanation in the sense of the term used here. So ontic accounts of explanation such as, for example, Craver's (2007) are not excluded by this approach.

2 Explanation in Interdisciplinary Contexts

The above preliminaries clarify my practice-oriented approach to scientific specialization and explanation. This section introduces the problem my project aims to solve. The trend of increased specialization motivates scientific research spanning multiple specializations—interdisciplinary research. Increasing prevalence of interdisciplinary research throughout the sciences is indicated by a number of empirical measures.[2] This trend is a response to increased specialization, which fragments scientific knowledge into narrow spheres of specialized technical expertise. Such fragmentation makes it increasingly likely that researchers from different specializations need to work together to answer significant questions about the world. This need is most obvious for practical challenges such as climate change, sustainable energy, and global health, which demand contributions from multiple scientific disciplines (Koskinen and Mäki 2016). But knowledge transcending the narrow confines of a single specialization is of epistemic value in many cases.

Its value notwithstanding, interdisciplinary research faces a number of challenges and obstacles. Many are institutional, such as funding opportunities, patterns of scientific work organization, peer review practices, expectations for promotion and tenure, and architectural design of workspaces. Such institutional challenges have been the focus of most research and policy efforts on interdisciplinarity to date (MacLeod 2018). However, interdisciplinary research also presents epistemic challenges, which stem from the difficulties of combining key aspects of the scientific method, such as data collection practices, modes of inference or analysis, norms of model construction, and standards for scientific knowledge. These challenges are a problem in social epistemology of science.

Explanation presents a particularly serious epistemic challenge for interdisciplinary research. As noted above, scientific specializations vary widely in their goals, methods, and epistemic standards. They also face stringent "selective pressures," in that a given specialization must continuously attract researchers and resources in order to persist. Specializations that do so tend to exhibit harmony or coherence in their goals, methods, norms, and objects of inquiry.[3] Such coherence is achieved over time and with hard-won labor, through the mutual adjustment of a specialization's objects of inquiry, goals, methods, and norms to one another. Concomitant with this mutual adjustment within a specialization is diversity across specializations. It follows that specializations are differentiated from one another in key aspects of the scientific method.[4] The epistemic challenges for interdisciplinary research are all aspects of the general problem of achieving harmonious fit among elements of a research program—methods, norms, and epistemic goals—that hail from two differentiated sources. Explanations are among the most distinctive epistemic products of a specialization.[5] Although the goal of explanation is broadly shared across the sciences, it is conceptualized and accomplished in diverse, locally specific ways. Different specializations impose different standards on construction and evaluation of explanations. Those standards, or norms, provide specific constraints on what a successful explanation should be like—the features it should exhibit, as such, and the ways it represents or otherwise relates to the target of explanation.[6]

The diversity of explanations across the sciences is mirrored by the variety of philosophical accounts of explanation, each of the alternatives being based on different scientific exemplars. For example, consider the Standard Model of particle physics, the linchpin of which is the Higgs mechanism.[7] A key part of the model is a gauge theory that unifies the known physical forces. It is constructed to satisfy two constraints: to retain the mathematical properties of gauge theory while breaking symmetry of the field. Empirical data constrain but do not determine the model; as many as 30 distinct models are known to be consistent with "the observable facts" in this case. The Higgs mechanism is, evidently, a significant scientific explanation. It is also highly distinctive in its relation to physical theory, unificatory aims, modeling constraints, and relation to empirical data. Extrapolating features of this explanation to other specializations imposes an external standard on the latter. For example, the double helix model of DNA structure, an exemplar of explanation for molecular biology, contrasts with the Higgs mechanism in all the respects noted above: it is not based on mathematical theory, does not aim to unify all biological phenomena, is constructed to accommodate a wide range of experimental results, and is interpreted causally or functionally. Researchers committed to one specialization's view of explanation often find the explanations of others wanting. The same goes for philosophers, who cannot as individuals be conversant with more than a few scientific specializations.

Although the problem has not been studied systematically, there is some reason to think that interdisciplinary projects across the sciences often fail to thrive, or even begin, due to discrepant explanatory norms and goals. Examples of failed interdisciplinary explanation have been documented in synthetic biology/humanities (Rabinow and Bennett 2012), prokaryote/eukaryote phylogeny (O'Malley 2013), and systems biology/stem cell research (Fagan 2016). If interdisciplinary researchers are committed to different views of explanation, this can lead to tension and forestall effective collaboration. Precisely because explanation is an important goal for many scientific specializations, yet understood in distinctive ways across specializations, it presents a stark challenge to interdisciplinary research. How, if at all, are genuinely interdisciplinary explanations possible?

Consider the simplest case, in which two independent specializations aim to explain the same phenomenon, such as the mechanical properties of solids under strain or human sexual behavior.[8] By "independent" I mean that the specializations involved are neither descendants of a recent split nor recently merged but which have been proceeding for some non-negligible time interval as separate lines of inquiry, constructing explanations in accordance with their distinct goals, norms, and methods. Once it is discovered that two independent specializations aim to explain the same real-world phenomenon, there are four possible outcomes: the two explanations may remain isolated from one another, coincide, conflict, or be integrated in some way.[9] The first is likely the most common, but is inapplicable to interdisciplinary research. The second outcome allows for smooth merging of distinct specializations into a new interdisciplinary specialization. That outcome is likely to be very rare, for the reasons noted above. Conflict, the third possible outcome, is the traditional response when different explanations of the same phenomenon are proposed—but like the first, this is incompatible with the goals of interdisciplinary research. So it is the fourth outcome that is of interest here. How can different explanations be integrated so as to construct a genuinely interdisciplinary explanation? This is the explanatory challenge for interdisciplinary research.

My strategy for answering this challenge is to weave together insights from the debate over scientific perspectivism and from the social epistemology of science. The former offers guidance in characterizing the possible relations between explanations constructed by different specializations. The latter indicates normative constraints that, applied to those relations, yield an account of explanation for interdisciplinary research. This result is not proposed as a general account of explanation but instead characterizes an epistemic virtue that is of interest beyond interdisciplinarity: comprehensiveness. This argument demonstrates that social norms for collaboration also have the epistemic benefit of increasing the comprehensiveness of explanation. That epistemic benefit can be understood in terms of perspectives as well. Combining distinct perspectives is a way of enhancing knowledge and understanding. That full account is beyond the scope of this chapter, however. The more modest task undertaken here is to show the relevance of the perspectivism debate for my project and to use this debate to ground a framework for analyzing relations between models constructed from different perspectives.

3 Models and Perspectives

Explanations constructed in interdisciplinary research may be conceptualized as models. Models have many functions in science, one of which is the explanation of phenomena of interest. The nature of explanation by models is an unsettled issue, with current debates focusing on whether representational accuracy is required (Frigg 2010; Kaplan and Craver 2011), whether such explanations must be causal (Woodward 2003; Rice 2015), and the role of similarity relations between explanatory models and their targets (Weisberg 2013; Parker 2015). The diversity of explanations across the sciences renders it unlikely that univocal answers on these points will be forthcoming. But one general claim about models in science does hold quite broadly: models and their roles are influenced by model users' purposes. If those purposes include explanation, then users' ideas about explanation inform their practices of model construction as well as resulting products. Methods of constructing explanatory models, and norms that guide these methods and constrain resultant products, vary widely across specializations. Explanatory models are, in this way, localized to the goals, methods, and standards of particular scientific specializations. In interdisciplinary research aimed at explanation, models or modeling resources from multiple specializations are needed to construct an explanatory model adequate to the purpose. So the explanatory challenge for interdisciplinary research can be reformulated in terms of divergent explanatory *models*, with the latter term referring to an epistemic product constructed for explanatory purposes by members of a specialization.[10]

The idea of explanatory models localized to scientific specializations is related to the notion of diverse perspectives. A perspective, in the everyday sense of the term, is a stance from which an object appears to one or more observers, roughly synonymous with "a point of view." From different perspectives the same object may present very different, even incompatible, appearances. The content of a perspective can be articulated as an indexical statement: "This is how it is from here." In art, the term has a more precise technical meaning, associated with strategies for realistically rendering three-dimensional scenes in two-dimensional media (Giere 2006; van Fraassen 2008). Visual perspective is characterized by having an origin, orientation, grain, marginal distortion, occlusion, and spatial distortion. The origin, traditionally, is the painter's or viewer's eye, while orientation is the direction in which that eye is looking—the angle of the gaze. Grain is the level of detail permitted by a perspective, ranging from rough to fine. "Marginal distortion" refers to the limits of the mode of viewing in a particular perspective. Occlusion is related; this term refers to focusing on some features so as to block or shadow others, which are thereby occluded in that perspective. Systematic spatial distortion is another hallmark of visual perspective: relative sizes of objects are represented differently than in reality (e.g., projection in two-dimensional maps of the earth).

These features of visual perspective have analogs in scientific modeling. Briefly, the origin corresponds to the user, orientation to features of the user's context, grain to the degree of abstraction, marginal distortion to the limits of the model, occlusion to the selection of some features of the target at the expense of others, and spatial distortion to idealization. Perspective in the everyday sense is also relevant to models in science. In philosophical discussions of scientific modeling, a model is taken to offer a (partial, often idealized) representation of a target object (Giere 1988; Morgan and Morrison 1999; Weisberg 2013). This partiality and contextual selectivity can be characterized as perspectival. For example, Mitchell and Gronenborn assert that, in scientific modeling, "what is represented and what is left out are usually tailored to meet some explanatory or pragmatic goal. This type of selection encodes what is sometimes referred to as a 'perspective'" (Mitchell and Gronenborn 2017, 707). Models of real-world targets do not, and arguably cannot, accurately represent those targets in every respect (Teller 2001). The features represented, and the degree of accuracy required, are chosen in light of model users' purposes. Such selective, partial representation is what makes models useful. In this sense, models are inherently perspectival. Their perspectival aspect permits multiple models of a single worldly target, incompatible with one another and yet all scientifically legitimate.

This situation is the starting point for debate about perspectival realism. But the perspectival view of models also applies to the explanatory challenge for interdisciplinary research. Different specializations produce, among their other results, explanatory models of phenomena of interest. Each partial, limited model is constructed by researchers within a specialization in accordance with its norms and methods for explanation. In this way, each specialization can be considered a distinct perspective within which explanatory models are constructed. The different specializations involved in interdisciplinary research contexts bring different perspectives to bear on the phenomenon of interest. Social scientists sometimes describe interdisciplinarity in exactly these terms. For example, Stichweh concludes his sociological study of scientific disciplines with the statement that "differentiation along disciplinary lines has the great advantage of viewing reality from radically different angles. The risks that the one-sidedness of each perspective entails are thus avoided" (1992, 12). In an important subclass of interdisciplinary contexts, the goal is to construct a single explanation from these diverse perspectives. This task is undertaken when specialized explanatory models are individually inadequate to account for the phenomenon of interest. In such cases, members of different specializations must somehow combine their respective explanatory models of the phenomenon of interest. So the challenge is one of integrating models constructed from distinct perspectives to create a single ("unified") explanation of the phenomenon of interest.[11] Responding to this challenge requires showing how this unification is possible, explicating and articulating conditions for its success, and distinguishing the resultant interdisciplinary explanations from other epistemic results. The first step is showing how models from different perspectives can possibly be integrated. And here the debate over perspectival realism offers important insights. That debate is not, of course, about explanation or interdisciplinarity; my approach here is to repurpose insights that emerge in that debate rather than contribute to it directly.[12] I do so by canvassing different relations between models from different perspectives that emerge in the debate initiated by Giere's (2006) *Scientific Perspectivism*.

In *Scientific Perspectivism*, Giere defends "perspectival realism" as a viable alternative to traditional scientific realism about theories, distinct from antirealist relativism. Giere proposes perspectivism based on the metaphor of color vision, arguing that scientific theories and observations (both models, according to Giere) are perspectival in the same way as descriptions of color: they tell us what the world appears to be like from a particular standpoint. All justified realist claims by scientists are therefore qualified and conditional. Giere takes these qualifiers and conditions to have the form: "according to this highly confirmed theory (or reliable instrument), the world seems to be roughly such and such" (5–6). A consequence of this perspectival account is that quite different descriptions of the same bit of the world are not incompatible. Rather, each describes a different way of experiencing the world, such that one is not an unqualifiedly better representation of that world than another. Disagreement (and normative selection of "the best" model) is possible only within a perspective.[13]

Giere's account contrasts with traditional scientific realism, which he characterizes in terms of the following conditional: if models constructed from different perspectives represent the world in incompatible ways, then at most one can be (approximately) true. It follows that in such cases there is direct conflict between models that must be resolved. The task then is to select, out of a set of competing alternatives, the model that is most likely correct. Traditional scientific realism thus allows for only one possible relation between distinct models of the same target phenomenon: *direct conflict* among competing alternatives. Giere rejects this traditional account, obviously, but does not otherwise say much about how models from different perspectives might relate to one another. Although he allows that agents can switch perspectives and compare results among them (83–84), Giere does not discuss combining or merging different perspectives, apart from the intriguing remark that an experimental test involves the "meshing" of theoretical and observational perspectives (89, 93).

In response to Giere, philosophers take a range of positions as to whether perspectivism is a genuinely realist alternative to both traditional scientific realism and to antirealism. In this debate the relation between models and the world is of primary concern, rather than between models from different perspectives. Yet ideas about this issue emerge obliquely, yielding an interesting array of results. Chakravartty (2010, 2017) argues, contra Giere, that perspectivism amounts to traditional scientific realism. Building on Rueger's (2005) proposal that incompatible models describe a system of interest relationally ("from this perspective it looks as if . . ."), Chakravartty argues that real-world systems have dispositions non-perspectivally and that these are differentially revealed by different detection methods (i.e., perspectives). Apparently incompatible models are reconciled by seeing them as revealing different dispositional facts about a single target system. An explanation, according to Chakravartty, describes how non-perspectival facts about a real system's structure produce the different behaviors/capacities that appear to us under different circumstances. This account of explanation posits an *indirect* relation between models from different perspectives; they are reconciled, or rendered consistent with one another, by an underlying structural description that coordinates them. Lacking a structural explanation, we can still answer contrastive "what-questions" by picking out the appropriate answer among the multiple perspectives on offer. Thus, another relation among models from different perspectives is via a long conjunction of the claims made by different models about a target T, stating each method and dispositions of the target revealed by that method. Such a conjunction has the form: "method x reveals a about T, and method y reveals b about T, and so on."[14] The relation between different models included in such a conjunction is minimal; one is linked to another via the formal connective "and," with no substantive relation posited to hold between any. I refer to this minimal connection between models from different perspectives as *simple additivity*.

Morrison (2011) takes issue with Giere's distinction between perspectivism and antirealism, arguing that the plethora of incompatible models of the atomic nucleus ascribe inconsistent "fundamental properties" to the target of inquiry, such that "there is no way to build on and extend the models in a cumulative way" (351). Construing alternative models of the atomic nucleus as offering different perspectives on its structure without the prospect of cumulative extension is, according to Morrison, tantamount to antirealism about atomic nuclear structure. Cumulative extension, she suggests, requires a unifying core or coherent treatment of all the different alternatives. The latter is available in other cases, such as models of turbulence, which are reconciled as different ways of elaborating the same set of basic principles; they are alternative idealizations based on a shared common structure. Morrison refers to this relation between models as "complementary," but as I argue below, that term is better reserved for a different relation. Morrison's core idea is much like Chakravartty's account of explanation: the different models do not directly relate to one another but are all subsumed by a more abstract or general model. Accordingly, I will refer to this relation as *subsumption*. Another possibility suggested by Morrison's account, although not one she discusses, is *direct cumulative interaction* among models from distinct perspectives.

Chirimuuta (2016) offers a different criticism of Giere, arguing that the metaphor of vision obscures some resources for perspectival realism and that the sense of touch, the operation of which involves active engagement and interaction with the world, is a more appropriate guiding metaphor. Her key claim is that "scientific representations inform us about the natural world in virtue of their interactive and interested qualities" (746). That is, scientific representations, although perspectival, can inform us about the world through interaction with it. On her "haptic" account, partiality, interestedness, and interaction are not obstacles to realism but constitutive of a realist role for perspectival representations in science.[15] Although Chirimuuta does not characterize relations between models from different perspectives per se, her account emphasizes the *interactive processual* nature of the model-world relation. Relations between models can be conceived in this way as well.

Massimi (2016a, 2016b) also looks to interaction to formulate an alternative to Giere's perspectival realism. She argues that a scientific claim meets the criterion of "success from within," supporting realist "success to truth" inferences, just in case that claim (i) performs adequately with respect to standards of its own, originating perspective; (ii) expresses a proposition that is in fact true; and (iii) meets standards of performance adequacy appropriate to its original context as assessed from another scientific perspective. The idea of (iii) is adapted from perspectival analyses of knowledge and belief, which distinguish context of use from context of assessment for knowledge claims. This contextualist distinction is adapted from that between a theorist's perspective and that of epistemic agent *S*, commonly invoked in analytic epistemology. In classic thought experiments, the theorist judges that *S* does (not) know that *p*, in accordance with standards that the theorist justifiably takes to be appropriate to *S*'s situation.[16] Perspectival theories of knowledge make these different stances part of the content of the theory itself. Massimi adapts the epistemological distinction between context of use and context of assessment to the case of multiple scientific perspectives. Scientists can judge the ongoing performance adequacy of a knowledge claim by the standards of its originating perspective, as those standards are interpreted in their own perspective. The criterion of success from within supports a realist distinction between approximately true parts of theories and "idle wheels" (after Kitcher). *Cross-perspectival assessment* is unlike the other relations canvassed here, in that it is not a relation between models as such, but a mode of relating different perspectives with respect to claims made in one. Cross-perspectival assessment tracks how every single perspective fares with respect to standards of performance adequacy when assessed from the point of view of other (synchronic or diachronic) perspectives.

Massimi's notion of a perspective as a "context of assessment" is compatible with the idea that specializations amount to different perspectives from which explanatory models are constructed. But an explanatory model typically consists of more than one claim (and often non-linguistic elements as well). If claims are considered to be parts of models, then cross-perspective assessment is an indirect relation, mediated by the standards of assessment constitutive of perspectives. In order for one perspective to assess the standards of performance adequacy of another perspective, the two must be sufficiently similar for the judgment to be warranted; conceptual distance blocks cross-perspectival assessment (Massimi 2016b). This is not a problem if different perspectives are conceived as stages within a historical lineage of, for example, theories of light, motion, or the atom.[17] But it does seem to block cross-perspectival assessment across disparate specializations, as the standards of performance-adequacy for explanation in particular are typically very different. I return to this issue in the conclusion.

This survey of positions in the recent debate over perspectival realism yields the following list of relations, which can in principle hold between models from different perspectives:

- Direct conflict (Giere 2006)
- Simple additivity (Chakravartty 2010)
- Subsumption; indirect reconciliation (Morrison 2011; Chakravartty 2010)
- Interactive process (Chirimuuta 2016)
- Cross-perspective assessment (Massimi 2018)
- Direct cumulative interaction (adapted from Morrison 2011).

The next task is to refine the above list of possible relations to articulate a conceptual framework for analyzing relations between models from different perspectives.

4 Cross-Perspective Relations Between Models: General Framework

Eliminating the relations inapplicable to interdisciplinary research and generalizing the others yields a preliminary taxonomy of possible relations between explanatory models from different specializations (Table 2.1).[18] The taxonomy is framed by two crosscutting distinctions (direct/indirect and similar/different) and two continuous axes (degree of integration and stage of model construction). The last generalizes the insight of Chirimuuta's

Table 2.1 Four-Place Taxonomy of Relations Between Models

	direct	*indirect*
similar	overlap	subsumption simple additivity cross-perspective assessment
different	conflict LK-complementarity	contrastive N-complementarity

haptic metaphor: that relations between diverse models may hold, or not, at different stages of the model construction process, ranging from an initial sketch to an explanation that is complete according to the standards of its specialization. The degree-of-integration axis is anchored at one extreme by the relation of simple additivity. The latter involves no substantive connection between models from different perspectives; their contributions are simply strung together in a conjunction, making the connection between models one of bare logical consistency. Most interdisciplinary research aimed at explanation will require more substantive connections between models. However, there is no need to rule it out in principle.[19] The opposite extreme on this axis is mutual dependence, or reciprocity, between all elements of models from different perspectives. This latter is illustrated by Chirimuuta's concept of an interactive process, as well as van Fraassen's (2008) view of ongoing "joint construction" of theoretical and experimental perspectives.[20] These two axes provide a general graphical framework for characterizing relations between models from different perspectives in terms of stage of model construction and extent of integration.[21]

Alongside this framework, the two crosscutting distinctions yield a four-place taxonomy of relations between models. The direct/indirect distinction is readily seen from the initial list of relations in section 3. Conflict between models is a direct relation: different models make incompatible claims about the same target and thus directly contradict with one another. Subsumption, in contrast, is an indirect relation: different models are rendered compatible by each being subsumed by a more fundamental model (Morrison 2011) or underlying structure (Chakravartty 2010). Chirimuuta's haptic metaphor prioritizes direct engagement. The general point is that models from different perspectives can either relate to one another directly or via an additional mediator. The most familiar direct relation between models is one of overlap or coincidence; different models have certain features in common. Such points of coincidence, or zones of overlap, can serve as "joints" that connect models across perspectives. But for models constructed in different perspectives, differences are likely to outnumber similarities. This is the second distinction: models may relate to one another in virtue of their similarities or in virtue of their differences. The latter may seem counterintuitive. To better motivate it, I next argue for an addition to the list above: *complementarity.*

Direct conflict, at first glance, appears incompatible with explanation in interdisciplinary research. As noted above, the challenge for such cases is to integrate or unify explanatory models from different specializations; direct conflict would seem to be a non-starter.[22] Nonetheless, despite first appearances, this relation can hold between models from different perspectives. What is incompatible with interdisciplinary research—or, rather, with the class of such cases aiming at explanatory models constructed with contributions from all participating perspectives—is the way direct conflict is resolved through traditional theory choice. The traditional view is that, faced with a set of conflicting models, we must select the best for the purpose at hand. But when multiple perspectives are in play, there is no neutral stance from which "the best model" can be chosen. Such a stance is the "view from nowhere" that perspectivists repudiate. In interdisciplinary research, each perspective has something to contribute but cannot hope to provide a satisfactory explanation on its own. So, in these cases, conflicting models cannot relate to one another as mere competitors.

There is, however, another way to conceptualize the relation among directly conflicting models from different perspectives: as a form of complementarity. The idea is illustrated by occlusion in visual perspective (see section 3). Occlusion is a kind of representational limitation on model construction in the visual perspective: selecting one feature to represent rules out or otherwise compromises successful representation of other features of the same target. A feature that is blocked or distorted by representation of another in a model is occluded in that model. One model's occlusion can be another's successful representation—this is one reason multiple models of a phenomenon are sought in the first place. But models that complement one another in this way cannot, in general, simply be combined into a single coherent model of their shared target. They directly conflict by representing the target in mutually exclusive ways. This is an example of a direct difference relation between models; they relate to one another by contrast.[23] But models that represent a target in conflicting (mutually exclusive) ways can be related more abstractly, as complements or "negative images" of one another. To return to the visual metaphor: that which one model casts in shadow, the other presents in full light. I will term this relation N-complementarity (for "negative"). This immediately suggests another way in which models from different perspectives can relate to one another: namely, by complementing one another more directly. In this case there is no occlusion; different models just supply what is missing from one another's representations of a shared target. Such "lock and key" complementarity (LK-complementarity) is one form of mutual dependence among models from different perspectives. LK-complementarity is a direct relation: different models represent different features of the target, compensating for one another's partiality. N-complementarity is indirect; the properties that N-complementary models represent the target as having cannot be combined into a single coherent model. Such models can be integrated only via mediating relations, such as van Fraassen's "duality" (2008) or subsumption by another model. Escher's tessellations (designs that fill the picture plane) exhibit both kinds of complementarity: two different visual perspectives represented in a single frame, interlocking in a pattern of contrasting colors and perfectly complementary shapes (LK-complementarity), yet the harmonious whole is itself an illusion with which the viewer engages by interpretively switching between ground and figure. The continuity of shape from edge to center, and the blending of symmetry and contrast, produce the appearance of a single scene that is also two.

This establishes a second distinction among relations between cross-perspectival models: similarity/difference. The distinction concerns features of models in virtue of which they connect with one another. Models can connect through sameness: having features common to both. Or they can relate to one another through difference: a contrast leveraged into a positive link. Most philosophical discussions of model-integration focus on similarities across distinct models: shared structure, properties depicted, and so on. The lesson of Escher's work, and of other examples of complementarity across perspectives, is that models can be interestingly related in virtue of their dissimilarities as well.

The main result from this taxonomy is that models from different perspectives can be integrated in virtue of similarities or differences, either to one another directly or else via some additional mediator. These two crosscutting distinctions, together with the continuous axes (degree of integration and stage of model construction), provide a general comparative framework for analyzing relations between models constructed in different perspectives. Models may be related to one another by similarity—having features in common—just as they may relate to their targets in this way. If this similarity relation is direct, then the shared features are points of overlap or coincidence between models from distinct perspectives. These areas of coincidence serve as "joints," connecting different models into a more inclusive model. Alternatively, the similarity relation may be indirect, requiring a mediator. The traditional indirect similarity relation involves subsumption of disparate models by a more "fundamental" (or abstract, or general) model of their common features (Morrison 2011). The mediator in this case is the more fundamental model. Indirect relations via mediators of this sort involve hierarchical assumptions, with the abstract shared feature being privileged. This privileging is often associated with explanatory power. In this way, explanatory issues are implicated in perspectivism, although explanation is not the main focus of those debates. Direct relations among models do not require hierarchical assumptions of this sort. Models from different perspectives may also relate to one another through difference, either directly or indirectly. The former is epitomized by the seamless fitting-together of jigsaw puzzle pieces via complementarity in shape. Indirect relations of difference involve some mediating conceptual construct, such as a part-whole hierarchy relating different levels of description, or cause-effect relations representing elements of one model as causes of elements of another. These mediators also bear on ideas about explanation. This returns us to the issues raised at the start of this chapter concerning explanation and collaboration.

5 Conclusion

In this chapter, I have argued that recent debates about perspectivism have important parallels with the explanatory challenge for interdisciplinary research. Without imposing a priori ideas about the nature of explanation onto scientific practice, the epistemic products of distinct scientific specializations recognized "from within" as explanations can be conceptualized as models. Philosophical discussions of perspectivism take as their starting point the situation of multiple incompatible models of a single phenomenon, focusing on the implications of this situation for scientific realism. The explanatory challenge for interdisciplinary research, characterized in terms of models, begins with the situation of multiple models of a single phenomenon, constructed by different scientific specializations in accordance with the distinctive goals, norms, and methods of each. The question for interdisciplinary research is: how (if at all) can models from different specializations be combined so as to produce an explanatory model spanning multiple fields of expertise? Different specializations amount to different perspectives on the phenomenon of interest. The notion of perspective rests on a visual metaphor, which maps onto key aspects of scientific modeling. Construction of explanations is one kind of modeling, which differs in its particulars across specializations/perspectives. So the question becomes: how to integrate, or unify, explanatory models from different perspectives?

The first step toward answering this question is to examine the possible relations between models from different perspectives. This was done by canvassing positions in the perspectivism debate and generalizing from them (Table 2.1). The next step is to identify, within this array of possible relations, the subset that yield interdisciplinary explanations. Although that task is beyond the scope of this chapter, I will conclude by sketching the approach and some preliminary results. Constraints on relations between models from different perspectives relevant to the explanatory challenge for interdisciplinary research are provided by social epistemic norms for collaborative activity. Construction of explanatory models spanning multiple specializations is a kind of collaborative activity, and so norms for this kind of social action apply. I propose a set of such norms based on recent work in social epistemology of science and social action theory, modifying these for the special case at issue. The resulting set of norms indicates that relations between models constructed from different perspectives should be (among other things) limited and mutual. These results, I argue, cohere with perspectivist insights as well. Some indications of this coherence can be seen by re-examining Massimi's concept of cross-perspectival assessment, the outlier among relations canvassed above.

Massimi's account of cross-perspectival assessment includes at least three points that cohere with social epistemic norms for constructing interdisciplinary explanations. First, her account is premised on the idea that our current perspective is not privileged with respect to judging the truth of theories (or parts of theories). The corresponding attitude for interdisciplinary researchers is to conceive of explanations in one's own specialization as one partial viewpoint among others; no single specialization is prima facie explanatorily privileged over others. Humility about one's own perspective is the common, crucial point. Second, cross-perspectival assessment applies not to entire theories or models but to particular elements of them—individual scientific claims, in her terminology. This fits with Massimi's view that changes of perspective are not abrupt and wholesale but instead accomplished by gradual conceptual change by inquirers working within (while modifying) established intellectual traditions. Rather than changing perspectives, my account is concerned with integrating models. Yet the same lesson emerges, that integration of models from different perspectives need not be thoroughgoing. Integration requires only a limited connection, a sturdy bridge from which to traverse from one part of an interdisciplinary model to another. Third, for the cases Massimi considers, it is reasonable to suppose enough overlap across perspectives for a warranted assessment of the standards of one perspective by another to be possible. But this is not generally the case for explanatory models in interdisciplinary research. Perspectives conceived as distinct specializations, such as theoretical physics and molecular biology, have much less substantive overlap than those that share an intellectual lineage, whether as contemporaneous rivals or in historical succession. Yet some counterpart to Massimi's cross-perspectival relation seems applicable: a sort of "self-other" location. The counterpart would be locating one's own model in relation to a model constructed from another perspective. Following Massimi's lead, this would be assessment of one's own model in terms of its relation to another. Roughly speaking, the idea would be that users of a model in one specialized perspective can see how their model connects with the models of other specializations in interdisciplinary research. That is, there is a way to "travel" from one modeling perspective to the others via the connecting bridge that links those models.

There is, evidently, much more to do in clarifying these ideas and developing them into an account of explanation for interdisciplinary research—or, rather, an account of one explanatory virtue that is important in interdisciplinary research: comprehensiveness. What I hope to have accomplished here are the first steps toward such an account, showing the relevance of perspectivism to the explanatory challenge for interdisciplinary research, using a range of positions in the recent debate about perspectival realism to ground a general framework for analyzing relations between models from different perspectives, and indicating ways that further insights from perspectivism dovetail with a normative social epistemic account of explanation in interdisciplinary research.

Acknowledgments

Many thanks to Michela Massimi for the opportunity to contribute to this volume and for valuable comments on an earlier draft. Thanks also to Hanne Andersen, Anjan Chakravartty, Mazviita Chirimuuta, Haixin Dang, Steve Downes, Sara Green, Lilia Gurova, Sune Holm, Kareem Khalifa, Jim Lennox, Elijah Millgram, Sandra D. Mitchell, and Maria Serban for valuable comments and feedback. Funding was provided by the National Science Foundation (SES-1354515), the University of Utah College of Humanities, and a generous donation from the family of Sterling M. McMurrin.

Notes

1. Cf. Kuhn (2000, 250) on "a vast and still accelerating proliferation of specialties."
2. Porter and Rafols (2009) report a more than 50 percent increase in prevalence and complexity of co-authorship of scientific publications, as well as diversity of cited sources in article references, between 1975 and 2005. The American Academy of Arts and Sciences (2015) charts a steady increase in interdisciplinarity in PhD dissertations from 2003 to 2012. Social scientists Brint, Turk-Bicakci, Proctor, and Murphy (2009) found steady growth of interdisciplinary programs in US four-year colleges between 1975 and 2000. Scholars of interdisciplinarity distinguish between inter-, multi-, and transdisciplinarity. In this paper, I use the term "interdisciplinary" and its cognates in a generic sense, encompassing these diverse forms.
3. This is related to what Hacking (1992) refers to as "self-vindication" for laboratory sciences in particular. But the point is more general: a close fit between different elements of scientific practice in successful lines of inquiry has been noted at least since Kuhn (1962).
4. An exception may be specializations that have recently split, such as genetics and genomics. The latter, though distinguished from the former in core aims, rhetoric, and concepts, shares a great deal with genetics; institutionally, their separation is incomplete. Other specializations emerge at the interstices of others (e.g., molecular biology, cell biology) and thus have extensive overlap with these "parents" (Darden and Maull 1977).
5. This claim is the obverse of Woody's (2015) account of explanation as having the social function of binding together scientific communities.
6. The practice-oriented approach taken here thus entails pluralism about scientific explanation. This contrasts with the traditional philosophical aim of analyzing scientific explanation and/or understanding in general. A more modest version of that project can be pursued, however, via analysis of multiple epistemic virtues that are shared more broadly (though not universally) across specializations. Some of these virtues are familiar: simplicity, scope, accuracy. However, different specializations not only associate different collections of virtues with explanation but also interpret the same virtue quite differently.
7. This example is based on King (2018). Note that the Higgs mechanism is not a mechanism in the causal sense usually assumed by philosophers today. The Higgs mechanism is (at least prima facie) non-causal.
8. Generalization to cases involving three or more scientific contexts is straightforward.

9. The problem is actually more general than as described above, where the explanations to be integrated are more or less finished products. Integration of explanations can occur at any stage in the process of constructing those explanations—the challenge raised by divergent goals, methods, and standards for explanation remains.

10. Regarding ontic explanation, see remarks in section 1.

11. For now, I will remain neutral as to whether explanations in interdisciplinary contexts must be understood realistically.

12. There is, however, a structural analogy between the perspectivism debate and my account of explanation in interdisciplinary contexts. Both are concerned with different specifications of one general question: how can multiple diverse perspectives relate to one another so as to yield an epistemic product that is in some sense unified? For perspectival realists, the epistemic product is knowledge of the one real world; for myself, it is a single explanatory model constructed via interdisciplinary research.

13. Note that Giere uses the term "perspective" somewhat differently than the sense meant here. Giere's term refers to a hierarchy of models that includes data models, representational models, and theoretical models. His usage equates perspectives with sets of models that are integrated in use. Different scientific specializations, in contrast, comprise agents and resources for constructing models. Multiple perspectives (in Giere's sense) can occur within a specialization (a perspective, in my sense). This terminological difference does not affect the canvassing of relations between models from different perspectives. Thanks to Michela Massimi for pushing me to clarify this point.

14. For example: "What is water? It is something that dissolves salt in some circumstances and not in others; that behaves like a continuous medium in some circumstances and not in others; that is nourishing in some circumstances and not in others," etc. (Chakravartty 2010, 412).

15. "We can think of models as devices which aim to achieve a certain fit between a natural phenomenon, the human mind, and our particular purposes. Explanatory, predictive, and practical success are a matter of achieving the right kind of fit, not of the attainment of some God's eye view on the subject. There can be various ways to be successful (a plurality of perspectives), and sometimes the best way to achieve a good match between the natural phenomenon, our conceptual resources, and the tasks we have at hand, is through willful distortion" (Chirimuuta 2016, 22).

16. Significantly, the theorist is presumed to have a "richer informational state" than agent S, which "entitles" him or her to make the cross-perspective assessment of epistemic standards (Sosa 1991).

17. Massimi (2016b, 105) does allow for "synchronically rival" as well as historically sequential perspectives in her account. That she considers them "rivals" seems to exclude interdisciplinary research aimed at explanation. However, in new work Massimi explores the idea of complementarity among perspectives (personal communication).

18. An earlier version of these results appears in Fagan (2017).

19. Contrastive explanation, which reconciles apparently conflicting models of the same target by an explanatory division of labor, may be an example of simple additivity. On the contrastive (erotetic) account, explanatory models answer distinct questions about the phenomenon of interest.

20. Note that van Fraassen's (2008) concept of a perspective is not the same as the one proposed here, although the two overlap in important ways.

21. Empirical studies of interdisciplinary research (Shrum, Genuth, and Chompalov 2007; Gerson 2013; MacLeod and Nagatsu 2016; MacLeod 2018)

suggest that efforts at "wholesale" integration across specializations are often unsuccessful. More restricted connections, leaving most specialized practices unchanged, are associated with better outcomes.

22. Models that directly conflict with one another make contradictory or incompatible claims about the target phenomenon, which cannot at once be true. Outright incompatibility between models from different disciplinary perspectives is unusual, as different specializations are more likely to talk past one another than to clash directly. Direct conflict is thus more common among models within a specialization than across them.

23. Van Fraassen (2008, 315–316) suggests an abstract formalism for analyzing relations of this sort: "duality," an operation on partially ordered sets that generalizes the idea of a complement (in the set theory sense) or negation. The details of the formalism and van Fraassen's particular use of it can be set aside here.

References

American Academy of Arts and Sciences. 2015. *Humanities Indicators: The Interdisciplinary Humanities PhD*. Accessed June 27, 2018 https://humanitiesindicators. org/content/indicatordoc.aspx?i=10882.

Brint, S. G., Turk-Bicakci, L., Proctor, K., and Murphy, S. P. 2009. "Expanding the Social Frame of Knowledge: Interdisciplinary, Degree-Granting Fields in American Colleges and Universities, 1975–2000." *The Review of Higher Education* 32(2): 155–183.

Chakravartty, A. 2010. "Perspectivism, Inconsistent Models, and Contrastive Explanation." *Studies in History and Philosophy of Science Part A* 41(4): 405–412.

Chakravartty, A. 2017. *Scientific Ontology: Integrating Naturalized Metaphysics and Voluntarist Epistemology*. Oxford: Oxford University Press.

Chirimuuta, M. 2016. "Vision, Perspectivism, and Haptic Realism." *Philosophy of Science* 83(5): 746–756.

Craver, C. 2007. *Explaining the Brain: Mechanisms and the Mosaic Unity of Neuroscience*. Oxford: Oxford University Press.

Darden, L., and Maull, N. 1977. "Interfield Theories." *Philosophy of Science* 44(1): 43–64.

Fagan, M. B. 2016. "Stem Cells and Systems Models: Clashing Views of Explanation." *Synthese* 193(3): 873–907.

Fagan, M. B. 2017. "Explanation, Multiple Perspectives, and Unification." *Balkan Journal of Philosophy* 9(1): 19–34.

Frigg, R. 2010. "Models and Fiction." *Synthese* 172(2): 251–268.

Gerson, E. M. 2013. "Integration of Specialties: An Institutional and Organizational View." *Studies in History and Philosophy of Biological and Biomedical Sciences* 44(4): 515–524.

Giere, R. 1988. *Explaining Science: A Cognitive Approach*. Chicago: University of Chicago Press.

Giere, R. 2006. *Scientific Perspectivism*. Chicago: University of Chicago Press.

Hacking, I. 1992. "The Self-vindication of the Laboratory Sciences." In *Science as Practice and Culture*, edited by Pickering, A., 29–64. Chicago: University of Chicago Press.

Kaplan, D., and Craver, C. 2011. "The Explanatory Force of Dynamical and Mathematical Models in Neuroscience: A Mechanistic Perspective." *Philosophy of Science* 78(4): 601–627.

King, A. 2018. "Explanatory Models: A Framework for Instrumentalism." *Presentation at Models and Simulations 8*, University of South Carolina, March 16, 2018.

Koskinen, I., and Mäki, U. 2016. "Extra-Academic Transdisciplinarity and Scientific Pluralism: What Might They Learn From One Another?" *European Journal for Philosophy of Science* 6(3): 419–444.

Kuhn, T. S. 1962. *The Structure of Scientific Revolutions*. Chicago: University of Chicago Press.

Kuhn, T. S. 2000. *The Road Since Structure*. Chicago: University of Chicago Press.

MacLeod, M. 2018. "What Makes Interdisciplinarity Difficult? Some Consequences of Domain Specificity in Interdisciplinary Practice." *Synthese* 195(2): 697–720.

MacLeod, M., and Nagatsu, M. 2016. "Model Coupling in Resource Economics: Conditions for Effective Interdisciplinary Collaboration." *Philosophy of Science* 83(3): 412–433.

Massimi, M. 2016a. "Three Tales of Scientific Success." *Philosophy of Science* 83(5): 757–767.

Massimi, M. 2016b. "Bringing Real Realism Back Home: A Perspectival Slant." In *The Philosophy of Philip Kitcher*, edited by Crouch, M., and Pfeifer, J., 98–120. Oxford: Oxford University Press.

Massimi, M. 2018. "Four Kinds of Perspectival Truth." *Philosophy and Phenomenological Research* 96(2): 342–359.

Mitchell, S., and Gronenborn, A. 2017. "After Fifty Years, Why Are Protein X-Ray Crystallographers Still in Business?" *The British Journal for the Philosophy of Science* 68(3): 703–723.

Morgan, M., and Morrison, M., eds. 1999. *Models as Mediators*. Cambridge: Cambridge University Press.

Morrison, M. 2011. "One Phenomenon, Many Models: Inconsistency and Complementarity." *Studies in History and Philosophy of Science Part A* 42(2): 342–351.

O'Malley, M. 2013. "When Integration Fails: Prokaryote Phylogeny and the Tree of Life." *Studies in History and Philosophy of Biological and Biomedical Sciences* 44(4): 551–562.

Parker, W. 2015. "Getting (Even More) Serious About Similarity." *Biology and Philosophy* 30(2): 267–276.

Porter, A. L., and Rafols, I. 2009. "Is Science Becoming More Interdisciplinary? Measuring and Mapping Six Research Fields Over Time." *Scientometrics* 81(3): 719–745.

Rabinow, P., and Bennett, G. 2012. *Designing Human Practices: An Experiment With Synthetic Biology*. Chicago: University of Chicago Press.

Rice, C. 2015. "Moving Beyond Causes: Optimality Models and Scientific Explanation." *Noûs* 49(3): 589–615.

Rueger, A. 2005. "Perspectival Models and Theory Unification." *British Journal for the Philosophy of Science* 56: 579–594.

Shrum, W., Genuth, J., and Chompalov, I. 2007. *Structures of Scientific Collaboration*. Cambridge, MA: MIT Press.

Sosa, E. 1991. *Knowledge in Perspective: Selected Essays in Epistemology*. Cambridge: Cambridge University Press.

Stichweh, R. 1992. "The Sociology of Scientific Disciplines: On the Genesis and Stability of the Disciplinary Structure of Modern Science." *Science in Context* 5(1): 3–15.

Teller, P. 2001. "Twilight of the Perfect Model Model." *Erkenntnis* 55(3): 393–415.
van Fraassen, B. 2008. *Scientific Representation: Paradoxes of Perspective*. New York: Oxford University Press.
Weisberg, M. 2013. *Simulation and Similarity: Using Models to Understand the World*. New York: Oxford University Press.
Woodward, J. 2003. *Making Things Happen: A Theory of Causal Explanation*. Oxford: Oxford University Press.
Woody, A. 2015. "Re-Orienting Discussions of Scientific Explanation: A Functional Perspective." *Studies in History and Philosophy of Science Part A* 52: 79–87.

3 What Is Perspectivism, and Does It Count as Realism?

Paul Teller

1 Giere's Perspectivism

Broadly, perspectivism is the view that human knowledge is unavoidably from a particular perspective or vantage point. Such ideas have resonated throughout the history of philosophy, for example, in the work of Leibniz, Kant, and Nietzsche, and in the contemporary work of some philosophers, such as Putnam and Nagel. Nancy Cartwright (1983) and Ronald Giere (1985) launched a line of work that looks at science not as an enterprise of finding true laws but one of constructing inexact models. They did not use the metaphor of perspectives, but their approach has been deeply perspectivist. Models are always simplifications, different ones working well for different explanatory and practical objectives.

Giere (2006) explicitly casts this line of work in the perspectivist metaphor, urging also that this perspectivism should count as a form of scientific realism. He begins (chapter 2) with the illustration of color vision. We are trichromats, thereby having a "colored perspective" on the world that differs from those of di- and tetrachromats. Giere then argues (chapter 3) that this understanding generalizes to observation with scientific instruments. The "perspective" that governs a measurement with an instrument is the collection of assumptions about the functioning of the instrument and of the theory used in interpretation of the instrument's output. Why does this count as a "perspective"? A measurement outcome is only secure insofar as these assumptions obtain. In particular, the measurement outcome can be safely used only to the extent that these assumptions apply or can be taken for granted. In this sense, measurement outcomes cannot be detached from the perspective within which they were obtained. In addition, no measurement perspective is unique because each involves imperfect idealizations.[1] These conclusions apply equally to scientific observations made without instruments because of the commonplace that, to be useful, even observations made without instruments have to be theoretically interpreted.

Turning to theoretical claims (chapter 4), Giere argues that in formulating theories scientists "create perspectives within which to conceive of aspects of the world" (59). Following the case of observation with instruments, a theoretical perspective is constituted by the assumptions of a theory. For the same reasons that applied in the case of observation, all claims made within a theory are undetachable from the perspective constituted by the theory.

Natural questions arise about Giere's perspectivism, as I have so far summarized it. Often a claim made in one perspective/theory can be used in a context governed by a different perspective. Perspectives can be combined, but "[m]ultiplying perspectives does not eliminate perspectives" (92). The only sort of perspective-free claim that one can legitimately make is the conditional, with a description of the perspective as antecedent and measurement outcome or theoretical claim as consequent, and even such claims will be from some (presumably very broad) perspective.

Objection! One can use tables such as that of the National Institute of Standards and Technology to look up physical constants and use them just anywhere. Not really. Giere will say: to use these one must be skilled in their use, involving understanding of their theoretical meaning and the theoretical contexts in which they can be applied. Take the following example: the mass of an electron = $9.10938356 \times 10^{-31}$ kg. Theoretically, the rest mass of an object isn't even a constant. It varies with the "impact parameter" with which it is measured. One has to know whether and, if so, how to take such complications into account. One also has to have a theoretical understanding of the unit of measure, the kilogram.

Giere claims that, although all scientific claims are from within a perspective, such claims count as a kind of realism when the perspective in question and the way it has been used meet high standards of scientific practice. Why? All that Giere tells us is that such are "the most reliable conclusions any human enterprise can produce" (92). I think that the intended argument goes something like the following: Giere has acknowledged (chapter 1) that such conclusions are in part contingent on human needs, interests, and historical accidents, since these affect our fashioning of details of our theories. But the reliability of our theories is also very heavily contingent on "the world," that is, on circumstances utterly beyond our control. Moreover, scientific standards insure that when accepted, such perspectival claims are still very robust. They will stand up in a wide range of relevant circumstances. They can be combined with cognate perspectives to expand our store of (still always perspectival) knowledge. In these senses, conclusions that meet high scientific standards are "telling us something about the world." This is the best—really the only—kind of knowledge of the world we can have. This is the most one could ask of a realist attitude. I don't think Giere has more than this to offer and, yes, so far the argument is thin. Below I will very extensively enlarge on such considerations as supporting the label "realism."

2 Scientific Realism as Referential Realism

One consideration that, in combination with others, will favor perspectival realism will be that standard scientific realism fails. I will work with a "generic" formulation of standard scientific realism, following common elements of two standard sources (Psillos 1999, xix ff.; Chakravartty 2017, sec. 1.2).

In the first instance, a scientific realist wants to say that our "mature" theories get it right about the world, that is, that they offer important claims that are true. Recognizing that, broadly, even our best theories are simplifications or idealizations, the claim of truth is, brushing aside worries about what it is to be approximately true, qualified as approximate truth.[2] But to realists, well-established existence claims seem unproblematically and not just approximately true. Indeed, what would it mean for a claim of existence of a thing or kind of thing to be "close to the truth"? Either there is that thing, or kind of thing, or there is not! (This issue will come in for closer examination below.)

It is useful to reformulate this generic statement. The canonical example is the claim that there are atoms. Instead of claiming that there are atoms, one can reformulate this material mode statement in the formal mode and say that "atom" has a non-empty extension. Generalizing on this example yields what I call "referential realism":

> *Referential realism*: Many important terms in mature scientific theories have non-empty extensions, and the theories in which they occur, at least often, provide approximately true claims about the things in these extensions.

Standard scientific realism is or includes referential realism as a principal component.[3,4] The point of my reformulation is that it will make apparent options that are otherwise not apparent.

To see what is involved in referential realism, I will appeal to what I call "the tools of reference." First, there are referring terms that, when things go well, we take to have specific referents. "The Eiffel Tower" refers to the Eiffel Tower. "The standard kilogram" refers to the standard kilogram.[5] "Temperature" refers to the quantity temperature. Next, there are predicates (understood also to include kind terms and relational terms), which, when things go well, we take to have specific extensions: for "atom," the collection of all atoms; for "cell" (disambiguated to the intended biological kind), the collection of all cells; for "chair," the collection of all chairs. (I include the problematic example of "chair" to foreshadow the fundamental point soon to come.)

How do predicates get attached to their extensions? This cannot be directly, for then a predicate's extension could not change and we could not reason counterfactually about such extensions (what if there were more or different kinds of atoms . . .). Instead, for each predicate there must be some mediating consideration, what I will refer to as an extension-determining characteristic, to which the predicate is attached or with which it is associated, so that something is in the predicate's extension just in case it has that characteristic. These characteristics might be properties, qualities, attributes, kinds, conditions, and so forth. What I will need here is only that these characteristics determine what things go into an extension. Otherwise I needn't take a stand on just what these characteristics are or how they get attached to predicates. Thus, my characterization of the "tools of reference" covers a vast range of "theories of reference," all of which seek to characterize how the presumed tools of reference get attached to their referents and extensions. Finally, the package includes the applicability of strict identity. If a and b have referents, then there is always a fact of the matter, whether or not $a = b$ (recall Quine on this point: "No entity without identity!"). And for each extension of a predicate and each entity, there is a fact of the matter whether or not the entity is in the extension; equivalently, whether or not the entity has the characteristic associated with the predicate in question. Referential realism, as I intend it, is to be understood in terms of these referential tools.

Once the role of the tools of reference has been made explicit, we see that there are two ways in which some specific claim of realism could fail. Using the example of atoms, it could be that the term "atom" has been successfully attached to an extension-determining characteristic, but that extension is empty. "Magnetic monopoles" might be an example of such failure. Or the term may never have been successfully attached to any extension-determining characteristic. Note that this could be, not because there are no candidate characteristics to be attached, but because there are *too many* characteristics with failure to have picked out exactly one of them. We will see that, in the first instance, this is what appears to happen for "atom."

3 Referential Realism Fails

The underlying problem is that, given human limitations, the world is too complicated for our theoretical terms to get attached to any extension-determining characteristics.[6] The easy case is atoms. Are ions atoms? If yes, then bare nuclei would count as atoms. If no, then salt, held together by ionic bonds, is not composed of atoms. In the case of covalent bonds, quantum chemists start with the nuclei in a molecule as a "backbone" and then estimate properties of the "electron cloud" that envelops this backbone. Woody concludes (2010, 427) that when seen through such techniques, "any reliance upon the concept of the isolated atom in developing molecular wavefunctions was gone."

Borderline plasmas provide another example. In a plasma there is enough energy to completely dissociate electrons from nuclei. Presumably there are no atoms in a plasma. But as the temperature of a gas rises, approaching that for a plasma state, the process will be incomplete, and there will be a quantum mechanical superposition of atoms, incompletely ionized states in which atoms have lost some of their electrons, and states with nuclei completely stripped of electrons. In such a state, it will be indefinite just which parts count as atoms.

Note that what I have so far argued is not that there are no candidates for the needed extension-determining characteristic to function for the term "atom," but that there are too many. This makes room for the following response to the difficulty that I have raised for referential realism. Let us grant that we use "atom" loosely so that, strictly speaking, it does not have a well-characterized non-empty extension. But this loose use can easily be tidied up. What actually exist are things like ions, complexes of nuclei and electrons in covalently bonded molecules, nuclei and electrons in plasmas, and so on.

But when we get to the puzzles of quantum theory all of this becomes, at the very best, uncertain. Nuclei are composed of quarks and gluons, and functioning as they do in vast quantum mechanical superpositions makes it dubious whether "quark" and "gluon" have specific extensions at any one time. Electrons likewise coexist in superpositions with positrons in pair creation and annihilation.[7]

The case of atoms illustrates the kind of problem that I have in mind. But the problems are ubiquitous. The sorts of complications that arise for water chemistry will compromise specific extensions for molecules, at least in many cases. For things larger than molecules, there will be the same problems that arise for objects of perception that I will examine below. But for starters, just what is the extension of "chair" supposed to be? Though less extreme, the same sort of problem arises for predicates such as "cell."

The obvious retort to these worries is: so much the worse for the tools of reference! Bear in mind that "the tools of reference" is an umbrella notion that covers any in the family of accounts that work in terms of reference and extension-fixing characteristics, however "characteristics" is more specifically understood and these characteristics come to be associated with their predicates. The natural alternative is the family of "use" accounts, stemming from Wittgenstein's "the meaning of a word is its use in the language," according to which one must look to see how a term is used in a language community. Often such use accounts are distressingly non-specific.

What about cluster concept accounts? If the cluster in question is fixed, this is just another kind of characteristic and falls under the tools of reference umbrella. If the cluster is open-ended, somehow to be used at the discretion of language users, this will fall under the use family.

A second worry. Obviously, the tools of reference function for us some-how extremely successfully. So when I argue that the tools of reference fail to function as described, many readers will feel sure that something must have gone wrong.

No one gives up an account when problems are pointed out unless pre-sented with an alternative account that meets at least some of the difficul-ties and that enables one to see how the old account worked as well as it did. So I must respond to this worry by introducing my alternative. We use the tools of reference, with great success, by using them in idealized models or similar arrangements.[8] Speaking metaphorically, in the world described by an idealized model, the referring terms and predicates *do* have referents and non-empty extensions. And in fortunate cases, in the real world such models function for us extremely well. In certain respects, this makes my approach a member of the use family. But my approach differs entirely from familiar versions of use accounts. These are sup-posed to work by looking at the patterns of use in a language community. My account instead looks at whether and to what extent an idealized use of a referential term functions for us successfully in respects that we care about. This makes my approach also very non-specific in a certain sense, but in a way that, I submit, is appropriate. My way of thinking about use of referential terms gives the *form* of answers to questions about how the term works. More specific answers to questions about success of an idealized use have to be answered case by case by examining just how a specific idealization, or a family of idealizations, functions.

The argument for my alternative account will be that it makes sense of the successful functioning of our use of the tools of reference in the face of the difficulties that I have explained. In particular, my examples will illustrate how taking the tools of reference to function through their use in idealizations simply sidesteps the difficulties.

Now to address a third anticipated complaint: the problems I have enu-merated are just problems of imprecision, understood as ambiguity and especially vagueness.[9] Well, yes, completely agreed, except for the "just"! Characterization in terms of vagueness reformulates the problems, and usefully so, but does nothing to resolve them. We have no account of vague-ness that succeeds in telling us how to make sense of "vague" extensions of vague predicates or, transposing to the formal mode, how to provide truth conditions for vague statements, such as characterizing an object as a chair. This is not the place to summarize the failings (with respect to the problems relevant here) of all extent accounts of vagueness.[10]

4 The Alternative to Referential Realism, Instrumentalism, Also Fails

If referential realism fails, presumably the alternative is instrumentalism. At least among other things, theory functions as a guide to what to expect by way of the deliverances of perception. Instrumentalism, broadly conceived, is the view that the content of scientific theories is exhausted by such expectations. But instrumentalism has the presupposition that unaided perception does not have the failings that undermine referential realism for theoretical objects. We take ourselves plainly to see things around us—the objects of perception such as the Eiffel Tower, the standard kilogram, and the like. Likewise, we take ourselves plainly to see property instances that these objects of perception have: the pointer pointing to the numeral "5," the right side of the balance pan being lower than the left side, and the previously blue litmus paper now being red.

But instrumentalism fails. It fails because this presupposition fails, and for the same kind of reason that referential realism fails in the realm of the theoretical. The world is too complicated for perception to identify completely specific concrete objects of perception or the properties or characteristics that such objects of perception might have.

Starting with properties, I will illustrate with the well-known example of perception of colors. Naively, we take ourselves visually to detect colors of objects as intrinsic properties that they possess: the color of that scarf is bright red. Already in the early modern period it was appreciated that colors, and secondary qualities generally, are not intrinsic properties but a complex of the interaction between external objects and our perceptual systems. Today we know a great deal about the complexities of color vision. Our experience of things as colored is a complex process involving properties of the object perceived, the light reflected, extremely complex perceptual processing, and all kinds of environmental circumstances.[11]

What about primary qualities, shape, size, and duration? As noted by many from Berkeley onward and in the respects relevant here, primary qualities suffer vagaries similar to those suffered by the secondary qualities. A coin viewed from an angle casts an oval image on the retina but is still seen as round. All kinds of environmental clues (relative positions and sizes, textures, lighting, etc.) affect how we see relative locations, sizes, and shapes. Perceived temporal duration is highly context dependent, and for short intervals even perceived temporal ordering of events can be a construction of our perceptual processing.

Turning to objects, we take ourselves plainly to see objects in front of us: that stone, that tree, that chair, the standard kilogram bar, and so forth. But what is actually perceptually available to us? To illustrate, imagine you are driving toward the Eiffel Tower. You catch sight of it—you see the Eiffel Tower! But wait—a bit of the top is obscured by a cloud. And even with nothing obscuring your vision (from the visual perspective you hold) you only see one side of a limited number of parts of the tower. That you have seen (all, or even a part of) a completely specific referent of "the Eiffel Tower" looks to be some kind of inference—but, of course, no explicit conscious inference. Perceptual experience fills in enormously.[12]

What current perceptual science shows us is that our perceptual system puts together for us a visual experience as of some completely specific external object. Not only does the perceptual system fill in much detail with which we are not currently in visual contact, but just how we experience the target includes a great deal of tacit information about how the object would look from different angles, about the further experiences we would have were we to physically interact with the object, and so forth.[13]

How might we interpret these comments? Two alternatives are available. First, there is some completely specific referent of "the Eiffel Tower" and likewise for other perceived objects. Our perceptual system fashions for us a pretty detailed model of these objects, filling in a great deal that is not visually available. But owing to the complexity of things, these models are never in any sense complete and never, where they do specify, completely accurate. The second alternative is just like the first, but the assumed target of perception is problematized. The alternative I want to consider is not a modern referential idealism (or solipsism), that all we have are our perceptions and there is nothing otherwise independent of us. Rather this alternative agrees, to express it neutrally, that there are "Eiffel Tower phenomena," but the world is far too complex for there to be one specific referent of the phrase "the Eiffel Tower." Rather, thinking (and acting) in terms of a unique referent is a simplification, a kind of idealization, of a much more complex set of circumstances.

This second alternative should by now sound familiar. What's the argument for it in the case of objects of perception? As before, the problem is not that there are no candidate referents but that there are too many. Just what is to be included in the supposed specific referent of the first alternative? Spatially, there are so many specific physical objects that could be in question: With or without the concrete buttress holding up the legs? With or without bits of paint that have just about flaked off? With or without a bolt that has just come off? Just what gets included in the purported referent of "the Eiffel Tower"? For the standard kilogram bar, the problem is particularly acute. The bar is constantly losing and gaining tiny bits of matter, for example, when it is handled. Technicians struggle to take these into account, to get a stable standard for the kilogram, but never with complete success. Just what is the referent of "the standard kilogram bar"? Temporally, at just what point in time did the Eiffel Tower or the standard kilogram bar come into existence?

That there are specific objects of perception, referents for our terms for them, and properties that we take ourselves as perceiving them to have, are simplifying idealizations or models that nature puts together for us. The problem and the response to it are very much the same as for the purported referents of our theoretical terms. To provide an alternative to realism about theoretical objects, instrumentalism would have to be free of the difficulties for theoretical objects from which it was supposed to rescue us. But instrumentalism faces exactly the same underlying difficulties. The world is too complicated for us to have any exact knowledge of it. At both the level of theory and the world of perception, we know the world though the inexact representations (call them models, idealized characterizations, perceptions, etc.) that we fashion or that nature fashions for us.

By this point we have a more detailed characterization of and argument for Giere's perspectivism, as applying not just to things postulated by our theories but also at the level of perception. All human knowledge is inexact. All human knowledge is perspectival in the sense that our representations are always fashioned within one or another inexact representational scheme. In the case of theory, we generally have a range of complementary theoretical perspectives. In the case of perception, for the most part we are built to work with the perceptual perspective that nature has fashioned for us, which is nonetheless a perspective because it is highly inexact. It is also shown to be a perspective, in the current sense, by the fact that we can complement it with one or another theoretical perspective that we use in the science of perception.

I need to make two qualifications to my claim that *all* human knowledge is from one or another inexact perspective. Combinatorial facts, logic, and in general, finite mathematics provide a most plausible systematic exception. But as soon as we get to mathematics where the incompleteness results show that there are unintended interpretations, we have difficulties analogous to those for empirical knowledge. Second, it is to be emphasized that the ubiquity of perspectivism is not a logical or conceptual fact about human knowledge. It is entirely contingent because of the complexity of the world compared to our meager human epistemic capacities, abundantly illustrated by examples that I have cited and innumerable ones similar to them. Because it is contingent, the conclusion is not that this is what human knowledge has to be. The world is changing so fast that I will not hazard how this all might look 100, 50, possibly even 20 years from now. Rather, I take my job as an interpreter of the human epistemic enterprise to characterize the nature of human knowledge as it now exists or will be as long as it faces anything like the limitations that now constrain it. But does the perspectivism that I have argued for properly count as some kind of realism?

5 How to Think About Realism

What are our options? Referential realism, or any view of which it is a necessary condition, fails. Instrumentalism fails. I won't discuss other historically discarded views: idealism, sense data theories, and so on. The problem is that we need a fresh alternative. Perspectivism provides such an alternative. Perspectival knowledge is the best, really the only knowledge (with the qualifications mentioned) humans can, in practice, have. We could let it go at that. The perspectivism that I have described isn't referential realism. Legions of scientific realists have laid claim to the term "realism." Perspectivists could just let them have the term and move on. But it is also worth exploring why, just as Giere suggested, the label "realism" is appropriate for perspectivism.

To review, with just a little more detail, here is the earlier "generic" formulation of contemporary scientific realism, following Psillos (1999, xix) and Chakravartty (2017, sec. 1.2). First component: there is a mind-independent world that is the target of scientific (and perceptual) knowledge and understanding. Since most of what happens happens independently of what we think or wish, perspectivists eagerly embrace this assumption. Second component: scientific claims are to be taken literally, in particular as having truth values, and not to be reinterpreted as mere instruments for guiding our expectations about the perceptual. Since, as I have explained, instrumentalism is no rescue for the difficulties instrumentalists see in realism, perspectivists also enthusiastically endorse this condition. Third component: mature theories succeed in giving us true, or approximately true, statements about the world and the things in it. Scientific realists concede that, for the characterization of the properties of things, all scientific knowledge is, at best, "approximately true." But generally speaking, the position appears to be that, for much of mature science, the referential claims are true, full stop.[14] I have already detailed my reservations about referential success. In addition, we need some critical examination of how to understand "approximate truth."

The reservation about approximate truth is that it is context dependent and, in particular, interest dependent. This already follows from the circumstance that "approximate truth" is going to be a matter of degree. Close enough to the truth for what? That will depend on what our interests are, what is at stake. But the context dependence is much more pervasive.[15] To say that a false statement is approximately true comes to saying that what the statement describes is similar to what in fact obtains. But similarity is context dependent and, in particular, interest dependent. Any two things are similar in countless ways and dissimilar in countless others. To say similar without explicit qualification always has to be understood as similar in contextually determined respects and not in others. Given that approximate truth can be understood in terms of similarity between what is claimed in a statement and what a fully correct statement would present, approximate truth has the same contextuality that we see in claims of similarity. One and the same statement will be approximately true in some respects but not in others. Which respects are intended is usually set by context.

Acknowledging the contextuality of approximate truth has an unexpected implication: standard scientific realism is immediately transformed into a perspectivist view! If, like absolute truth, approximate truth were context independent, then one could say of an approximately true statement that it is approximately true for, well, for anything. But given that an approximately true statement will be (a) close to the truth in some respects and not in others and (b) in those respects close enough for some things but not others, these context-dependent limitations constitute restrictions to perspectives in the sense developed in section 1. Specifying the respects and degrees already is robustly to characterize a perspective from which, or within which, a statement can be treated as true, a perspective from which claims cannot be exported.

Proper treatment of the existence claims of referential realism injects a second robust perspectivist component into standard scientific realism. For the existence claims themselves, "approximate truth" may seem to many a non-starter. What would it mean to say the claim that there are atoms is "approximately true"? Either there are atoms or there are not! I don't know of any explicit statement to this effect in the realism literature, but context often suggests that this attitude is being presupposed, perhaps as so obvious that it needs no mention. The one hedge I know in the literature is Psillos's statement: "[T]he entities posited by [mature, well confirmed theories], *or, at any rate, entities very similar to those posited*, do inhabit the world" (1999, xix, emphasis added).

How should we understand saying that entities of one kind are "very similar to" entities of another kind? If both kinds of entities exist, such claims are unproblematic (as long as we keep in mind the complication that two objects that are similar in some respects will always be dissimilar in others). But if there are no entities of the first kind, and this is because the term for the first kind hasn't been attached to an extension-determining characteristic, it is without content to say that entities of a second kind are similar to those of the first kind. No entities of the first kind have been specified. So there are no things of the first kind of which we can say that they share properties with entities of the second kind. The best we could say is that entities of the second kind have the (or many of the) properties that entities of the first kind were *supposed* to have. But then, why would they—namely, the entities of the second kind—not be the real entities of the first kind to begin with? So Psillos's hedge is empty. We need an alternative way of addressing the referential failure of standard scientific realism.

Though having rejected the existence claims of referential realism, I have also recognized the usefulness of the tools of reference by thinking of them as operating in idealized models (accounts, theories, etc.) where they function as if they had referents, and the models as a whole, in fortunate cases, function excellently in understanding one or another general aspect of how the world works. But understanding the function of the tools of reference as working through application in idealized models or idealization schemes is to characterize them as operating within the perspective characterized or created by those idealizations. The upshot is that, once the difficulties with both approximate truth and the referential component have been addressed, standard scientific realism *becomes* a perspectivist account.

Still, one could insist: so much the worse for realism of any kind for science. And, for those who have been persuaded by my arguments for the world of perception, so much the worse for realism for the objects of perception also. I think this would be hasty. While rejecting application of the designation "realism" in any absolute non-perspectival account, I will offer some considerations that support the appropriateness of using the term for the perspectival account I have sketched.

If we think of "realism" as denoting an attitude that takes imperfect representations nonetheless as telling us about the mind-independent world, this is an attitude that current scientific realists should endorse. By and large, "approximate truth" is good enough for realists to say that a theory is telling us a lot about how the world is and, one can add redundantly, how it is *really*. Let me offer a model, or parable, that supports this attitude.

Imagine a much simpler universe. You and I know every detail of what occurs in this world. Denizens of this world get a great deal pretty accurately, but they get nothing exactly right. They can give descriptions of how macroscopic objects are shaped but get some little details wrong. The small mistakes don't show that they have no grip on what such shapes *really* are, especially when the mistakes are, for their interests, inconsequential. Or consider a thermodynamic example. This world is composed of Democritean-like particles. There is no further analysis to be had of their constitution. These particles engage only in rectilinear motion that can change on collision. The creatures of this world postulate a quantity, temperature. They have thermodynamics but no statistical mechanics. We know that their temperature is mean translational kinetic energy for which there are fluctuation phenomena, but the world's denizens aren't aware of these fluctuations and incorrectly believe that there are none. Still, they know a lot about thermal phenomena. Their "temperature" is an idealized quantity that applies to their world only through their idealized thermodynamics. Still, these denizens know a lot about what is really going on in their world, especially if the fluctuation phenomena are inconsequential to them.[16]

I submit that such thinking shows "realism" to be a sensible label for "approximately true" qualitative claims, and if so, the label is sensibly extended to existential claims likewise. The barrier to so doing was worries about how to understand "approximate truth" for claims such as "things of kind x exist." But I have supplied a viable interpretation. To say that "things of kind x exist" is approximately true is to say that an idealized model that uses the tools of reference for things called "x" is successful for an important and broad range of applications.

I will develop two further ways to support the appropriateness of retaining the designations "real" and "realism." I suspect that a facet of the problems with the realism debate has been roughshod use of the word "real." In the first instance, "real" is used for the simple idea of reference. To say that the Eiffel Tower is real is just to say that "the Eiffel Tower" has a referent. To say that Santa Claus is not real is just to say that "Santa Claus" has no referent. To say that "atoms are real" is just to say that "atom" has a non-empty extension. But then all sorts of more subtle but important uses and connotations of the word "real" are left by the wayside. Are chairs, clouds, rainbows, shadows, holes, mirages, and fairies real? There is a slide here, where there shouldn't be, if the world "real" is applied with the clear "all or nothing" presumption. No question: chairs are real. Clouds also, I should think. Are rainbows real? Any discomfort with saying that rainbows are real is that they are not real physical objects. You're in trouble if you go looking for the end of one. But rainbows are perfectly real optical phenomena. Similarly for shadows and holes. They are perfectly real, but one has to be careful what one means by such statements. (Real) shadows are relatively well-localized blocked light in an otherwise well-illuminated background. (Real) holes are the relatively well-localized absence of material in an otherwise solid medium. Turning to mirages, they, like rainbows, are perfectly real optical phenomena, but we also deny that they are real in the sense that there are no objects of which they are mirages. And things pretty well give out when we get to fairies.

The example of clouds nicely illustrates the way I want to think about how we use the tools of reference in a way that shows the foregoing slide to be unproblematic. We do use the tools of reference with clouds as referents and extensions. We talk about "the oval-shaped cloud" and "all the clouds in the sky at noon." And identity: "you and I are looking at the same cloud." But *of course* we have to take care with such usage. We use the tools of reference when the purported referents are "distinct enough." When clouds begin to merge such usage breaks down. Treating clouds as specific, distinct objects of reference is a simplification—an idealization, if you will. The "distinct enough" is easily understood as follows: this idealization is one that does not get us into trouble with our present practical aims. When our subject is clouds, the standards of care and the things that can go wrong are different from and much more demanding than the analogous standards of care when the subject is ordinary physical objects. When we get to optical phenomena, such as rainbows and mirages, we can still use the tools of reference, but with their own required standards of care. The standards of care are dependent on the kind of idealizations that are in question for these different subject matters—the idealizations that are needed to enable use of the tools of reference.

The moral is thus the following: scientific realists have proceeded as if concrete particulars and other purported referents, such as, perhaps, space-time points, property instances and the like, and their collections were all that is in question. But when using "real," one has to take care to be clear: real for what kind of things? Concrete particulars? Phenomena? Absence of light or material? And so on. Then we have to look at whether, or to what extent, treating the referent as real—more carefully, using referential terms in the simplification—will work for us in the present practical context. It is the strategy of deploying the tools of reference, and other simplifications, in one or another broad idealizing scheme that Giere had in mind with his characterization of working within one or another perspective.

Let me provide a second consideration that illuminates this general moral. Readers who have followed me this far won't succumb to the following, but some already in section 4 will have had the following reaction: the author is arguing that there are no colors, rocks, chairs, cells, electrons, and so forth. This is just rank skepticism! Some of these critics may think that skepticism about cells and atoms is arguable, but hardly for colors and rocks.

But wait! Does sophistication about color perception amount to skepticism about colors? In a sense, yes. But such skepticism is utterly harmless. Once one sees that such skepticism about colors is innocuous, one quickly sees that the same goes for rocks and other macroscopic objects. One's mind is then opened to taking a similar approach to "unobservable" things such as cells and electrons, especially when one sees that instrumentalism provides no refuge.

To summarize the point: there is a kind of harmless skepticism that pervades the argument of this paper. But it is entirely mitigated once one takes a little care in how to understand the designation "real."

Interlocutors may object at this point: if we really are in a situation like that of the creatures of the world of my analogy, of course, the designation "realism" is well grounded. But take away what you and I know about in the story—"the way things *really* are, *exactly*"—and the analogy collapses. This worry gets my intent wrong. Nothing I have argued, claimed, or said involves or presupposes that there isn't some unique way things really are, exactly. "Things are what and the way they are and not some other way" is a truism to which I heartily subscribe. Perspectivism is a manifestation of our limited epistemic powers, relative to the overwhelming complexity of the world. With the qualifications mentioned above, exact human knowledge is utterly beyond our reach. We are always operating within some partial and not completely exact representational scheme. Different schemes attain their level of success differentially with different aspects of things. It is these representational schemes that correspond to what Giere had in mind by "perspectives." They are, in fortunate cases, immensely informative perspectives on the way things are. Add redundantly, if you like, the way things are *really*.

Notes

1. Throughout this chapter I use "idealization" very broadly for assumptions known to be false but advanced in the expectation that their errors will not interfere with the idealization's usefulness.
2. Except for entrenched instrumentalists, will any of us deny the truth of, for example, "water is H_2O"? The problem here is that both "water" and "is

H_2O" are imprecise or vague. The complexities of water chemistry show that as soon as the claim is sharpened up, it becomes false. See VandeWall (2007). This sort of complication applies broadly across the sciences. In Teller (2017), I work out in detail how such considerations impact claims of truth.

3. The referential component of scientific realism is often mentioned (e.g., Psillos 1999, xix; Chakravartty 2017, sec. 1.2; Laudan 1981, 33). There is also a critical literature (Hardin and Rosenberg 1982; Cruse and Papineau 2002; Newman 2005; Papineau 2010).

4. Does referential realism cover structural realism? I do not have the space to go into this issue in detail, but on the face of it, yes, since for a structural realist our theories must have terms interpretable as referring to the claimed structures.

5. Which I am told will have been kept and maintained after the new theoretical-based definition of the kilogram goes into effect sometime during 2019.

6. Occasionally critics demand argument that the world is indeed so complicated. Examples in this section and throughout the chapter will provide many illustrations.

7. While it requires more careful examination than I have so far given it, complications of electroweak theory also may throw extensions for elections and other leptons into doubt.

8. It is misleading to characterize such an approach as "fictional." That a description is partially in error does not make it a piece of fiction. Consider, for example, psychiatric case histories that falsify irrelevant personal details to preserve privacy. For a full discussion, see Teller (2009) and especially Winsberg (2009).

9. I have examined the interconnection between ambiguity and vagueness in this particular network of problems in Teller (2018a, 293–294). The well-understood phenomenon of ambiguity largely falls out of the discussion and I will focus on vagueness in the immediately following.

10. I have detailed some of these considerations in Teller (2018b). Also to be noted: since the problem I have described can be reformulated as problems of vagueness, the approach to addressing them, in terms of idealized descriptions, provides the seeds for an approach to vagueness that is entirely different from anything I know of in the literature. While I have not yet developed this idea in any detail, I have developed tools that I expect will be needed in Teller (2017).

11. For an accessible survey of the science, see Giere (2006, chap. 2). Chirimuuta (2015, chap. 4) gives a more detailed exposition.

12. Skeptics about what I am claiming here should see Churchland and Ramachandran (1998).

13. There are hundreds of research articles supporting these claims. Just three are Akselrod, Herzog, and Öğmen (2014); Churchland, Ramachandran, and Sejnowski (1994); and Churchland and Ramachandran (1998).

14. The references in note 3 above can be extensively extended.

15. The following is a rough and ready exposition to facilitate seeing what the problem is. A much more careful exposition is to be found in Teller (2001, 402–406). As explained there, the problem of similarity for approximate truth is the right interpretation of Miller's infamous "language dependence" problem.

16. In Teller (2018a), I show that *our* notion of temperature, and physical quantities very broadly, have exactly the idealized status illustrated in the foregoing analogy.

References

Akselrod, M., Herzog, M. H., and Öğmen, H. 2014. "Tracing Path-Guided Apparent Motion in Human Primary Visual Cortex V1." *Scientific Reports* 4: 6063.

Cartwright, N. 1983. *How the Laws of Physics Lie*. Oxford: Oxford University Press.

Chakravartty, A. 2017. *Scientific Realism. Stanford Encyclopedia of Philosophy*. https://plato.stanford.edu/entries/scientific-realism.

Chirimuuta, M. 2015. *Outside Color: Perceptual Science and the Puzzle of Color in Philosophy*. Cambridge, MA: MIT Press.

Churchland, P. S., and Ramachandran, V. S. 1998. "Filling in: Why Dennett Is Wrong." In *Critical Essays, 1987–1997*, edited by Churchland, P. M., and Churchland, P. S., 177–203. Cambridge, MA: MIT Press.

Churchland, P. S., Ramachandran, V. S., and Sejnowski, T. J. 1994. "A Critique of Pure Vision." In *Large-Scale Neuronal Theories of the Brain*, edited by Koch, C., and Lewis, J. I., 23–60. Cambridge, MA: MIT Press.

Cruse, P., and Papineau, D. 2002. "Scientific Realism Without Reference." In *The Problem of Realism*, edited by Marsonet, M., 174–189. Aldershot: Ashgate.

Giere, R. 1985. *Explaining Science: A Cognitive Approach*. Chicago: Chicago University Press.

Giere, R. 2006. *Scientific Perspectivism*. Chicago: University of Chicago Press.

Hardin, C. L., and Rosenberg, A. 1982. "In Defense of Convergent Realism." *Philosophy of Science* 49(4): 604–615.

Laudan, L. 1981. "A Confutation of Convergent Realism." *Philosophy of Science* 48(1): 19–49.

Newman, M. 2005. "Ramsey Sentence Realism as an Answer to the Pessimistic Meta-Induction." *Philosophy of Science* 72(5): 1373–1384.

Papineau, D. 2010. "Realism, Ramsey Sentences and the Pessimistic Meta-Induction." *Studies in History and Philosophy of Science* 41(4): 375–385.

Psillos, S. 1999. *Scientific Realism: How Science Tracks Truth*. London: Routledge.

Teller, P. 2009. "Fictions, Fictionalization, and Truth in Science." In *Fictions in Science: Philosophical Essays on Modeling and Idealization*, edited by Suárez, M., 235–247. New York: Routledge.

Teller, P. 2017. "Modeling Truth." *Philosophia* 45(1): 143–161.

Teller, P. 2018a. "Measurement Accuracy Realism." In *The Experimental Side of Modeling*, edited by Peschard, I., and van Fraassen, B. Minneapolis: University of Minnesota Press.

Teller, P. 2018b. "Making Worlds with Symbols." *Synthese*. https://doi.org/10.1007/s11229-018-1811-y.

VandeWall, H. 2007. "Why Water Is Not H_2O, and Other Critiques of Essentialist Ontology From the Philosophy of Chemistry." *Philosophy of Science* 4(5): 906–919.

Winsberg, E. 2009. "A Function for Fictions: Expanding the Scope of Science." In *Fictions in Science: Philosophical Essays on Modeling and Idealization*, edited by Suárez, M., 197–189. New York: Routledge.

Woody, A. I. 2010. "Concept Amalgamation and Representation in Quantum Chemistry." In *Handbook of the Philosophy of Science. Volume 6: Philosophy of Chemistry*, edited by Hendry, R. F., Needham, P., and Woody, A. I., 405–442. Amsterdam: Elsevier.

4 Realism and Explanatory Perspectives

Juha Saatsi

1 Introduction

A realist portrayal of science should accommodate the fact that science describes the world from numerous "perspectives." The nature of these perspectives and their interrelationships have for long been the bread and butter of history and philosophy of science. Realists continue grappling with challenges arising from the contingencies of grand theoretical perspectives (or "paradigms") and the ever-increasing plurality of models used for predicting and explaining various phenomena. These challenges turn on the inconsistencies that science seems to harbor, threatening the realist credo that science successfully works by virtue of "getting things right about the world." A natural realist hope is that these inconsistencies can be accommodated through an apt notion of "perspective," which is compatible with the basic realist credo.

What notion of "scientific perspective" should realism incorporate then? Answering this question helps with understanding scientific realism, and it is further instigated by recently developed perspectivist foils to more traditional realism by Ronald Giere, Paul Teller, and Michela Massimi.[1] These self-proclaimed "perspectival realists" have developed and defended views about the *perspectival nature of scientific knowledge* that put emphasis on it "being situated" in historical and modeling contexts (Massimi 2018b, 164). Thus perspectivists characterize scientific knowledge as "the inevitable product of the historical period to which those scientific representations, modeling practices, data gathering, and scientific theories belong," and as being embedded in "a prevailing cultural tradition in which those scientific representations, modeling practices, data gathering, and scientific theories were formulated" (Massimi 2018b, 164).

I am doubtful that the realist's optimism and commitment toward scientific progress and theorizing (especially in the fundamental sciences) are best captured in terms of scientific knowledge. Articulating an alternative vision for realism is a book-length project. The limited aim of this chapter is to present a different, realist-friendly notion of scientific perspective that does not concern knowledge. In particular, I wish to focus on what I call "explanatory perspectives" in relation to a (minimally) realist commitment to accumulating scientific *understanding*. Shifting the focus from knowledge to understanding yields a different kind of perspectivism, since understanding (as presently explicated) is not knowledge but rather an ability. The factivity of knowledge (namely that knowledge entails truth) is an almost universally accepted platitude about knowledge. By contrast, I will argue that explanatory perspectives in science and their indispensability spring specifically from the *non-factive* aspects of theoretical representations that maximize our scientific understanding.

In particular, I will argue that insofar as scientists' understanding can be enhanced by idealizations and/or false metaphysical presuppositions—whether mistakenly believed or merely entertained as useful fiction—such non-factive aspects of theorizing naturally give rise to mutually incompatible perspectives on natural phenomena. Briefly put, explanatory perspectives are ways of thinking about and representing a subject matter (say, light) in an explanatory context, which function to augment our understanding of the natural phenomena we are theorizing about (say, the rainbow). We will see, furthermore, that understanding-enhancing non-factive aspects can be involved in theories that best support genuine explanatory understanding in a given historical or modeling context, without scientists necessarily knowing all the respects in which these theories are idealized or false. All in all, I will conclude that increasing scientific understanding does not just amount to accumulating knowledge, since understanding is not factive in the way knowledge is, and non-factive aspects of theoretical representations can increase our understanding without us knowing their non-factive status. Rather, what matters for explanatory progress is that understanding-providing theories and models de facto latch on to reality in appropriate ways so as to satisfy explanations' basic factivity requirement (to be explicated below).[2]

The present focus on explanatory understanding is limited, of course, but not unprincipled. Taking a stance on scientific explanations, and the kind of understanding they provide of natural phenomena, is critical for demarcating realist commitments, since realists typically take scientific explanations seriously in a way that antirealists do not. For realists, "explanation" is a success term: the mind-independent reality determines whether scientists have actually succeeded in explaining and providing genuine understanding. To this end, realists defend a suitably *factive* conception of scientific explanation: genuine explanations must "latch on to" explanatory features of the unobservable reality.

The realist conception of explanations' factivity must be immediately qualified. On the one hand, the assumption that (genuine) explanations are in some sense factive is an integral part of the realist stance toward scientific reasoning and its progress. On the other hand, clearly, scientific explanations do not require "truth and nothing but the truth," for otherwise none of our current theories or models (which invariably incorporate falsehoods, approximations, and idealizations) would count as explanatory. The burgeoning literature on scientific explanation contains various suggestions for how to understand explanations' (qualified) factivity. I will begin by sketching one idea below, based on the counterfactual-dependence account of explanation (section 2). Regarding the issue of perspectivism more specifically, I will then argue that the ensuing account of explanatory understanding allows the realist to identify, accommodate, and motivate various perspectival aspects of science. The argument is based on a study of different theoretical perspectives from which optical phenomena have been explained and understood. Focusing on various explanations of the rainbow, I will show how a realist commitment to steady progress in scientific understanding is compatible with the fact that it has involved numerous mutually incompatible metaphysical perspectives on light (sections 3 and 4). In conclusion, I will reflect on the "realist" content of the ensuing perspectivism about explanatory understanding (section 5).

2 Explanatory Understanding and Perspectives

A realist portrayal of explanatory understanding is best painted with a clear conception of scientific explanation in mind. It is hard to make sense of how explanations "latch on to" reality unless we begin with a sufficiently clear account of what explanations *are* and how they *work* (Saatsi 2018b). To this end, I will now sketch one account discussed in detail elsewhere.[3] The key idea of this *counterfactual-dependence account* is that explaining is a matter of providing information about systematic patterns of counterfactual dependence. Explanatory counterfactuals are appropriately directed and change-relating, capturing objective, mind-independent modal connections in the world that show how the explanandum depends on the explanans. The explanandum and the explanans, conceptualized as variables that can take different values, stand for suitably individuated worldly features. Explanatory counterfactuals provide what if things had been different information, indicating how the explanandum would have been different had the explanans been different. Paradigmatic explanation-supporting relations are causal, but the counterfactual-dependence account also applies to various kinds of non-causal explanations, which appeal to geometrical, mathematical, or non-causal nomological dependencies based on, for example, symmetries.[4]

If explaining is a matter of providing information that correctly answers *what-if* questions, it is natural to regard as more powerful those explanations that enable us to answer *more* such questions (with respect to a given explanandum). This simple idea has rich implications regarding the notion of explanatory power (or "depth"), since there are many ways in which explanations can be compared regarding their potential to enable us to answer more or less such questions. Detailed analyses of explanatory power along these lines have been provided by, for example, Hitchcock and Woodward (2003) and Ylikoski and Kuorikoski (2010). The latter distinguish four aspects of explanatory power:[5]

> "**Non-sensitivity**" stands for an explanatory generality, having to do with the range of values that the explanans variables can take without breaking the explanatory relationship.

> "**Precision**" stands for the degree of precision in which the explanandum is individuated relative to some contrast class.

> "**Degree of integration**" stands for the connectedness of an explanation to other theoretical frameworks as a means of extending the range of *what-if* questions that an agent can (more easily) answer with respect to particular explananda, for example, by virtue of equipping the agent with new inferential resources.

> "**Cognitive salience**" stands for "the ease with which the reasoning behind the explanation can be followed, how easily the implications of the explanation can be seen and how easy it is to evaluate the scope of the explanation and identify possible defeaters or caveats."
> (Ylikoski and Kuorikoski 2010, 215)

Explaining is a distinctive human activity, the goal of which is the provision of *explanatory understanding*, which we can think, along with Ylikoski and Kuorikoski (2010), as an ability to answer correctly a range of *what-if* questions in relation to a given explanandum. The more such answers an agent is able to provide (by an appropriate measure), the better understanding she has. In the light of this conception of understanding, there are both epistemic and pragmatic dimensions to explanatory achievements and progress of science. While the counterfactual-dependence account is a broadly speaking realist one (assuming an appropriate reading of the modalities it involves), the way in which explanations provide understanding requires that human beings stand in an appropriate cognitive relationship to them. It is a realist account by virtue of incorporating the *basic factivity requirement* that explanatoriness primarily derives from explanation latching on to worldly things that bear an objective, explanatorily relevant dependence relation to the explanandum. But explanatory theories and models also typically involve non-factive aspects that have to do with the pragmatic, human-related dimension of understanding. This is due to the way in which explanatory power can in various ways be increased by allowing a degree of misrepresentation in an explanatory theory or model.

Two of these ways are particularly pertinent to us. First, information about explanatory dependence can often be conveyed more effectively by using a representation that idealizes either the target phenomenon or the explanatory dependence at stake. The simplifying falsehoods that idealizations incorporate can thus contribute to an explanation's cognitive salience, and/or its degree of integration, and/or its non-sensitivity (Ylikoski and Kuorikoski 2010). Second, information about explanatory dependence can be most effectively grasped through a non-veridical metaphysical image of the system at stake. For instance, in many theoretical contexts human beings find it easier to cognitively operate in terms that are more familiar and concrete. Even if these cognitive benefits are brought about through partially misrepresenting the target or conceptualizing it in a wrong way—for example, in the way that *fluid* models of energy and electricity do (de Regt and Gijsbers 2017, 70–71)—they can help to provide genuine understanding, to the extent they enable theorists to correctly answer *what-if* questions that are underwritten by relevant explanatory dependencies in the world. (For example, one can use a fluid model to efficiently grasp dependences between electric current, resistance, and voltage.)

For a quick illustration, consider a simple explanatory model of tides as a sine function mapped on to the relative positions of the moon and the sun. Although the real explanatory dependence is not exactly sinusoidal, considerable mathematical convenience and cognitive salience (for anyone familiar with sine functions) is gained by modeling it as sinusoidal. Similarly, representing the gravitational effect of the sun and the moon in terms of Newtonian gravitational force ("pulling" the water) can enhance this explanation's cognitive salience (in a typical explanatory context), despite misrepresenting gravity as a force (acting at a distance). Modeling tidal phenomena in these terms can provide a powerful explanation, tracking the dependence of tides on the explanans variables (namely relative positions of the moon and the sun) accurately enough, in a way that enables an agent (with suitable training) to answer numerous *what-if* questions regarding the explanans.

This simple example illustrates the interplay between explanations' factive and non-factive aspects in providing explanatory understanding. On the one hand, tides really do counterfactually depend on the relative positions of the sun and the moon; the explanation is factive and explanatory to the extent it captures this dependence. On the other hand, an idealized representation, with non-veridical metaphysical posits to boot, can provide better understanding than a more faithful representation by virtue of enabling us to better answer more *what-if* questions, by making the dependence of tides on the explanans variables cognitively more salient to us. In this way the "user-friendliness" of an explanatory theory or model can trump fidelity as an explanatory virtue, since what matters is the understanding that it provides limited cognitive beings like us with particular inferential skills and training. Recognizing the importance of cognitive salience also helps to appreciate how the factivity requirement leaves room for the possibility that maximal explanatory understanding is effectively gained from several mutually incompatible theoretical contexts. For instance, while some *what-if* questions regarding tides can only be correctly answered in the context of general theory of relativity (with curved space-time and no gravitational force), the various *what-if* questions that arise in, for example, oceanography are best answered in the context of Newtonian gravity in a way that involves gravitational force.[6]

I will argue below that this kind of interplay between factive and non-factive aspects of explanations accounts for how different "explanatory perspectives" naturally arise in science. To anticipate the discussion of the rainbow below, consider 19th-century wave theorists of light, who advanced scientific understanding from the perspective of various ether theories. Going further back, the likes of Descartes and Newton presumably also advanced scientific understanding of light from their respective theoretical perspectives. More synchronically, in the contemporary context we can regard geometric ray and electromagnetic wave models of light, along with the models of modern quantum optics, as offering complementary perspectives on the whys and hows of light phenomena. These different theories and models have steadily advanced the scientific understanding of light, I will argue, by virtue of providing accumulating information about the dependence of light phenomena on various features of the world. These explanatory features are captured by explanans variables upon which the explanandum phenomenon depends in a way that is quantitatively encapsulated in these theories and models. This accumulation of factive explanatory content is compatible with radical differences in these theories' and models' ontologies and metaphysical presuppositions, which need not be factive. These (often) non-factive presuppositions can nevertheless form a cognitively indispensable part of the theoretical context in which the explanations are situated, as we will see below in relation to various explanations of the rainbow.

As a scientific realist, I wish to maintain that advances in scientific understanding are achievements that relate to the way the world is beyond the observable phenomena. Here is an obvious challenge: how to explicate the sense in which Descartes, Newton, Fresnel, and others advanced *genuine* explanatory understanding of light given that their explanations presupposed mistaken explanatory posits (e.g., elastic ether). Is it not the case that their explanatory successes were merely *apparent*, undermined by the subsequent ontological shifts away from their mistaken explanatory posits? In response, some philosophers forgo the factivity assumption (and realism), construing "explanatory understanding" so as to allow them to maintain that past scientists achieved genuine understanding despite their radically mistaken theories (de Regt and Gijsbers 2017; de Regt 2017). In the realist spirit, I am inclined to insist that genuine understanding requires factivity with respect to the relevant explanatory dependencies; hence I will respond to the question above by explicating this factivity in a way that is compatible with past theorists' understanding of light being irretrievably entwined with their particular theoretical and metaphysical perspectives. Luckily, the counterfactual-dependence framework provides a way to do this by virtue of allowing factive explanations to naturally incorporate also non-factive aspects that are broadly pragmatic and contextual.

From this point of view, theories and models that are false in various ways and degrees can provide genuine explanatory understanding by underwriting theorists' ability to make correct what-if-things-had-been-different inferences. To the extent these inferences are furthermore made true by (causal or non-causal) dependence relations in the world, a theory or model latches on to reality in a way that fulfills its explanatory function regardless of its non-veridical aspects. Moreover, these explanatory counterfactuals can invoke explanans and explanandum variables that relate to unobservable features of reality, giving sufficient substance to realist commitment regarding explanatory understanding.[7] So while the non-factive, pragmatic dimension of explanations, involving idealizations and metaphysical presuppositions, can give rise to different explanatory perspectives, one's realist commitment need only concern explanations' factive dimension and the progress that science de facto makes with respect to it (regardless of whether or not scientists *know* which aspects of their explanations are factive).

3 Reflections and Refractions on Explanatory Perspectives

Different explanations of the rainbow illustrate well realist commitment toward accumulating scientific understanding. From the dawn of science, the rainbow has challenged scientists, primarily as an object of explanation (as opposed to experimentation or intervention).[8] Various explanations of (different aspects of) the rainbow have been provided by generations of physicists, including many of the most illustrious minds in the history of science. These explanations have been provided from varied theoretical and metaphysical perspectives, spanning different scientific paradigms and modeling practices. Nevertheless, we will be able discern a steadily accumulating factive backbone of scientific understanding that transcends the radical shifts in the changing perspectives, from Descartes, through Newton and ether theorists like Fresnel, to the modern day. From the viewpoint of the counterfactual-dependence account, we can view the radical shifts in the metaphysics of light, which have motivated antirealist arguments from the history of science (Laudan 1981), as being part of the non-factive aspects of these explanations. This account thus enables the realist to explicate the sense in which there has been genuine accumulation of scientific understanding of the rainbow from Descartes onward.[9]

What does it take to "explain the rainbow"? Like any typical physical phenomena, there are various *aspects* of the phenomenon that can be singled out as the explanandum, as reflected by the following questions.

1. Why does a rainbow have the shape it does?
2. Why does the (primary) rainbow form an angle of approximately 42° from the antisolar point?[10]
3. Why do we see a secondary rainbow at approximately 51° from the antisolar point?
4. Why is there a darker (Alexander's) band of sky between the primary and the secondary rainbow?
5. Why does the primary rainbow have the color pattern it does (red on the outside rim, violet on the inside)?
6. Why does the secondary rainbow have the color pattern it does (red on the inside, violet on the outside)?
7. Why are there smaller "supernumerary arcs" occasionally visible inside the primary rainbow, with a specific spacing between them?

René Descartes conducted a detailed study of the rainbow, and published explanations of (1) through (4) in *Discours sur la méthode* (1637). According to Descartes, these aspects of the rainbow can be explained in terms of the spherical shape of the raindrops in combination with a refraction of light (into a raindrop), internal reflection, and a further refraction (out of a raindrop). By using a combination of graphical analysis and numerical calculations to trace the geometry of light rays, Descartes discovered that these assumptions about light and rain give rise to a higher concentration of light at the scattering angle of 138° for a single internal reflection and 129° for two internal reflections (corresponding to 42° and 51° angles of the primary and secondary bows from the antisolar point, respectively). Furthermore, the fact that no ray involving one internal reflection can be deflected less than 138°, and no ray involving two reflections can be deflected more than 129°, can be related to the relative darkness of Alexander's band.

Descartes's explanations were provided from within his "modificationist" theory of light, according to which our perception of colors is due to the way in which light's transmission rotates otherwise stationary ether particles, the variable spin of which causes our sensation of different colors. Needless to say, this metaphysics is radically at odds with our physics. For example, since Descartes assumed light's transmission to be *instantaneous*, it was not possible for him to think of this transmission as unfolding *over time*, involving refraction, a subsequent internal reflection, followed by a further refraction. Another metaphysical presupposition of Descartes's theory was that the law of refraction was due to light traveling *faster* in a denser medium (e.g., water) than it does in air (Dales 1973). (Many have puzzled over the consistency of this presupposition with Descartes's assumption that the speed of light is not finite!) Such vastly mistaken metaphysical notions and non-referential terms involved in Descartes's theorizing might seem to render his explanation of the rainbow wholly surpassed by later theories, and unsuitable as an object of realist commitment of any sort.

This would be hasty, however. The realist can side with the standard historical narrative, according to which Descartes was the first to gain understanding of several important features of the rainbow. In essence, this is because the features of light relevant for Descartes's geometrical analysis are entirely continuous with high-school-level geometrical ray optics, namely, the law of reflection and Snell's law of refraction. We can further explicate Descartes's understanding and its factivity from the viewpoint of the counterfactual-dependence account. Descartes managed to explain (1) and (2) by virtue of grasping the way in which the rainbow's apparent location (relative to the location of the light source and the observer) depends on the shape of the raindrops and the density of water (responsible for the specific angle of refraction).[11] By virtue of getting these dependencies right, Descartes gained the ability to correctly answer various *what-if* questions. For example, he would have been able to work out how things would be different if the reflecting drops were made of glass instead of water.[12] To the extent he gained this ability, Descartes had genuine understanding of the rainbow. The historical fact that he wasn't able to theorize and express the relevant dependencies independently of his overarching mechanistic worldview and metaphysics of light rays does not nullify this understanding.

Notably, Descartes was altogether unable to account for the colors of the rainbow. Newton's advance is standardly taken to consist in realizing that the index of refraction (e.g., for water) is different for different colors, and that white light from the sun is in some sense a "combination" of many colors. These critical ideas of the color-variability of refraction allowed Newton to answer questions (5) and (6). These ideas are, of course, again embedded in Newton's broader corpuscular theory of light, according to which light is composed of non-spherical particles, with red corresponding to the larger and more massive particles than those corresponding to blue, for instance. Mechanical laws involving corpuscles' motion through luminiferous ether would account for the law of refraction in terms of differences of velocity in different media. (In Newton's "emissionist" theory denser media, such as water, "pulled" these corpuscles differently depending on their size and mass, resulting in a higher velocity component perpendicular to the interface.) Again, the broader perspective within which Newton's explanation was embedded is well off the mark on the whole, but a realist can nevertheless maintain the standard story according to which Newton genuinely advanced scientific understanding of the rainbow. From the viewpoint of the counterfactual-dependence account this advance can be explicated in terms of the *further explanatory dependences* that Newton got right, involving a new explanatory variable corresponding to the color of light and a dependence of the angle of refraction upon that variable.[13] The key to Newton's explanatory advance is an approximately correct *quantitative* representation of this dependence.[14] This enabled Newton to calculate the widths of the primary and secondary rainbow, for example, and it enabled him to answer new *what-if* questions about rainbows. For example, unlike Descartes, Newton was in a position to consider how these widths would be different if the drops were made of more or less dispersive medium. Similarly, Newton and his followers explicitly worked out how tertiary (and higher-order) rainbows would appear, were the light intense enough to give rise to them (Boyer 1959, 247).

The Newtonian account still leaves some directly observable features of the rainbow unexplained. In particular, it says nothing about the *supernumerary arcs* that can occasionally be seen inside the primary rainbow (and sometimes also on the outside of the secondary bow). An explanation of these supernumeraries requires the introduction of new explanatory variables that go beyond geometrical ray optics that Newtonian corpuscular theory exemplified. These variables can be found in the wave theory of light, which encompasses optical interference phenomena responsible for the supernumeraries. Thomas Young first realized that the spherical shape of raindrops makes it possible for there to be two ray paths with different angles of incidence (into the drop), internally reflected at the same point at the drop's rear surface, such that their final angle of refraction is the same. For light of an appropriate wavelength this gives rise to destructive and constructive interference, resulting in the supernumerary arcs. This theoretical treatment renders the drop size (relative to the wavelength of light) a new explanatory variable upon which the spacing of the supernumeraries depends. Furthermore, Young's interference theory of the rainbow explained also a number of other puzzling qualitative features that had been observed. For example, it explained why the bow is brighter near the earth and why the supernumerary arcs usually only appear near the highest part of the bow: these features depend on the relative size of raindrops, which tend to increase in size as they fall.

Again, these advances in scientific understanding were embedded within a particular broader perspective on the nature of light: Young (at the time in question) not only adhered to an optical fluid ether theory but also regarded light waves as longitudinal, akin to sound. This early wave theory was radically mistaken in many ways and unable to account for, for example, the polarization of light, but it nevertheless encompassed the right explanatory dependencies between the relevant explanatory variables, which are carried over to the later theoretical perspectives of the elastic solid ether theory, as well as the electromagnetic theory and beyond.

The subsequent idea that light waves were transverse was developed in a mathematically sophisticated way by Fresnel to explain various polarization phenomena. This now provided understanding of features of the rainbow that aren't visible by the naked eye, such as the fact (first noted by Biot in 1811) that the rainbow light is strongly polarized in the tangential direction.[15] Again, this explanatory advance was embedded within Fresnel's broader elastic ether theory of light. Since such ether does not exist, prominent antirealists have hailed Fresnel's theory an exemplar of a highly successful theory that is not even approximately true, undermining (certain kinds of) "convergent" realism (Laudan 1981).[16] Be the status of Fresnel's theory as "approximately true" as it may, the realist can stand by the standard story that takes Fresnel's contribution to explanatory understanding of the rainbow to be both genuine and lasting: the new explanatory variables introduced by Fresnel's explanations (e.g., light's wavelength relative to the drop size and the direction of light's polarization) capture further explanatory dependencies in the world. The historical fact that Fresnel (and his contemporaries) were unable to express and theorize about the relevant explanatory dependencies independently of the metaphysics of elastic ether does not nullify this contribution.

A realist would, of course, expect the theoretical perspectives on light subsequent to Fresnel to also recognize and build upon the explanatory dependencies that his theory captures. As far as I can see, this expectation is fully borne out in the rich history of accumulating understanding of the rainbow that continues still today. For example, over the last couple of decades there have been advances in understanding further aspects of meteorological rainbows in terms of their dependence on the distribution of *non-spherical* (oblate) raindrops (see Haußmann 2016 for a review).[17] The shape of raindrops has thus become an explanans variable in a deeper, more concrete way than it was before.[18] Furthermore, typical rain showers feature a broad variety of different drop sizes. It is an outstanding (although already partly met) challenge to work out how different observable features of the rainbow (e.g., colorization or the exact shape or brightness distribution) depend on new explanans variables that quantify a rain shower's physical features, such as its drop-size distribution and the drops' deviation from perfectly spherical shape.

These challenges largely belong to the domain of applied mathematics, a solid basis to which is provided by an exact description, in terms of Maxwell's electromagnetic theory, of the scattering of plane wave from a transparent dielectric homogeneous sphere, provided by Lorentz (in 1890) and Mie (in 1908). In the next section I will briefly discuss some developments in this area of applied mathematics, but I have already said enough to outline a realist stance toward the progressive trend that started with Descartes and has continued ever since. In the realist spirit we can view science as providing genuine understanding of natural phenomena, such as the rainbow, in terms of features of reality "behind the appearances." This presupposes a conception of explanation and understanding that is *factive* (albeit in an immediately qualified sense), supported by the counterfactual-dependence account of explanation. This account allows us to explicate the accumulating understanding in terms of scientists' increasing ability to answer counterfactual *what-if* questions regarding various explanatory variables. Our theories and models capture better and better how different explanandum variables depend on different explanans variables. These variables capture the dependence of different aspects of the rainbow on physical features of the world that are not observable, such as the raindrops' shape and their size relative to light's wavelength, and the direction of light's propagation and polarization. The accumulation of this factive content is fully compatible with the fact that different explanatory theories and models also have non-factive elements that give rise to mutually incompatible perspectives on light, due to, for example, the different ontological and metaphysical presuppositions that were an inextricable part of Descartes's, Newton's, and Fresnel's theorizing about light.

4   Which Explanation Is the "Best"?

So far, I have looked at the accumulation of understanding over the history of changing "paradigms" in optics. Let's now consider the (minimal) realist outlook in relation to mutually incompatible models employed in the current state of the art. The classic Lorentz-Mie theory of scattering can be regarded as the "complete and fundamental" theory of rainbow scattering. It is taken to deductively entail all the optical properties of an "ideal" rainbow.[19] Since this model of Maxwell's theory contains all the answers to different *what-if* questions about the (ideal) rainbow, one might think that we have reached the explanatory bedrock (regarding "ideal" rainbows)—the ultimate explanatory framework. Yet understanding of the rainbow has progressed much further since the inception of the Lorentz-Mie theory. Since scientists regard the subsequent development of, for example, idealized "semi-classical" explanatory models to provide deeper understanding, a realist must acknowledge the indispensability of further explanatory perspectives beyond the "complete and fundamental" theory. Hence, in some sense the fundamental theory only provides a limited explanatory perspective, which needs to be complemented by other vantage points to yield more comprehensive understanding. How should a realist interpret this plurality of explanatory models?

Different explanatory perspectives at stake here can again be understood from the viewpoint of the counterfactual-dependence account. In order to explicate the explanatory value of the idealized "semi-classical" models, I first need to say a few words about these further advances on the Lorentz-Mie theory.[20] These advances primarily turn on approximation schemes, such as the Complex-Angular-Momentum (CAM) method, which aim to extract the key features of the dynamics of the electromagnetic wave in a way that makes them transparent to us. As Nussenzveig puts it:

> A vast amount of information on the diffraction effects that we want to study lies buried within the Mie solution. In order to understand and to obtain physical insight into these effects . . . it is necessary to extract this information in a "sufficiently simple form."
>
> (Nussenzveig 1992, 45)

This simplicity, which is "to some extent . . . in the eye of the beholder" (Nussenzveig 1992, 210), can be achieved by suitable "semi-classical" approximations, occupying the rich theoretical borderland between geometrical ray theory and the wave theory. By working with idealized ray-theoretic concepts, while simultaneously making sufficient room for interference and diffraction effects, these approximations yield theoretical representations that render the relevant explanatory dependencies cognitively more transparent.

Although the Lorentz-Mie theory provides an exact solution of plane wave scattering by ideal spherical drops, it has the pragmatic downside of leading to a mathematical series that converges very slowly for particles of the size of raindrops. Thus, this theory is *oracular*: a powerful enough computer can crunch through a sufficient number of terms (typically several thousands) to yield however precise values of scattering amplitudes one desires, against which approximate solutions can be compared. However, due to the high number of terms and the series' lack of further physically interpretable structure, it provides no insight into aspects of the scattering process upon which the spacing of supernumerary bows depends. (A Laplacean demon might disagree, of course!) The first step beyond the Lorentz-Mie theory is to shift to the Debye series, which mathematically decomposes the wave front into "partial" waves, some of which are externally reflected, some transmitted directly through the drop, and some transmitted after n internal reflections. This series, which also provides an exact solution (equivalent to the Mie series), captures at the level of the wave theory the idea that the overall scattering dynamics can be represented as a sum of different processes, involving, for example, light that undergoes one internal reflection before transmission (responsible for the primary bow), light that undergoes two internal reflections (responsible for the secondary bow), and so on, with some of the light being "trapped" inside the drop for a number of revolutions before transmitting. However, the Debye series by itself does not allow us to identify which aspects of the scattering dynamics thus represented critically contribute to the features of the supernumerary bow.

Enter the CAM method. This approach allows the slowly converging partial wave series to be transformed into an approximate, rapidly converging expression in terms of "poles" and "saddle-points" in a complex-valued angular momentum space, representing the main contributions to the scattering amplitude at the primary rainbow angle. An interpretation of these poles and saddle-points in terms of both wave theoretic concepts (e.g., "tunneling" and "evanescent waves" near the drop's surface), *as well as* ray-theoretic concepts, provides the best means to bring out those aspects of the overall scattering process upon which the explanandum depends. By doing so it improves our explanatory understanding of the supernumeraries. Thus our best understanding of the rainbow involves representing light both as a wave and as a ray. How should a realist understand this plurality of incompatible perspectives? On the face of it, the explanatory indispensability of ray concepts could be taken to suggest that the ray-theoretic perspective is revealing features of light scattering that the wave theoretic perspective somehow misses.

I think the counterfactual-dependence account nicely captures the explanatory power of the semi-classical CAM perspective, even if we take Maxwell's theory to provide the "fundamental" story.[21] This is due to the importance of explanations' cognitive salience (cf. section 3). To illustrate this, consider a specific explanandum: why is the spacing S of the supernumeraries of a given rainbow 1.65°? From the counterfactual-dependence viewpoint, an agent understands the spacing if she is in a position to answer *what-if* questions of the sort "how would S be different if . . ." with respect to explanans variables, that is, wavelength and drop size, over some range of possibilities. Using the Lorentz-Mie theory the agent is capable of answering these questions, but only if assisted by a sufficiently powerful computer. The way in which the explanandum depends on the explanans is cognitively opaque to her.[22] The CAM approach provides deeper understanding by virtue of enhancing the agent's ability to answer such questions by revealing a much simpler explanatory dependence of the scattering amplitude on the explanans variables, without compromising the level of accuracy required for answering the explanandum at stake. This simplicity is not just an increase in computational efficiency but also a matter of representationally breaking down, in an idealized way, the overall Mie scattering into distinct processes, only some of which effectively contribute to the rainbow by largely determining S as a function of the explanans variables. This explanatory dependence is cognitively more transparent to us, and hence a theory that captures it provides (in a sense) a better explanation. In this way the counterfactual-dependence framework explicates the progress in the understanding of the rainbow achieved by moving from the *exact* Lorentz-Mie theory to the CAM *approximation*, the less fundamental explanatory notions of which (such as light rays and evanescent "surface" waves) can thus feature in our "best" explanation of the rainbow. This improvement is not a matter of introducing new explanatory variables that ontologically transcend the Lorentz-Mie theory (Pincock 2011), nor is it a matter of providing more fine-grained information about the explanatory dependence. Rather, the

improvement has to do with the way in which the CAM approach identifies critical explanans variables upon which the explanandum depends *in
a simple way*. These variables and the explanatory dependencies are fully
grounded in the wavelike nature of light and its dynamics; they are not
indicative of properties that somehow transcend Maxwell's theory.

What counts as the "best" explanation partly depends on the context
that determines how the different dimensions of explanatory depth are
weighed. The CAM approach can be taken to provide the most powerful explanation in the context of "pen and paper" mathematical physics,
while in the context of a computer-assisted study of actual (non-ideal)
meteorological rainbows, with variable-sized hamburger-bun-shaped
drops, the generalized Lorentz-Mie theory backs the most powerful
explanatory understanding.[23] There is no objective answer as to which
explanation is the "best" independently of such contextual factors. By the
same token, even though the earlier explanatory accounts from Descartes
onward are all strictly speaking false (even if we ignore their supereroga-
tory metaphysical content), these accounts can still be valuable sources
of understanding, and they can indeed be viewed as providing the "best"
explanation of certain aspects of the rainbow in suitable explanatory
contexts. For example, in the context of high school physics the gist of
Descartes's account (sans Cartesian metaphysics) provides the best explanation, simply because it provides the cognitively most transparent way
to capture the dependence of the approximate angles of primary and secondary bows upon the spherical geometry of raindrops given the laws of
reflection and refraction. Overall, the indispensable plurality of (strictly
speaking) incompatible explanatory perspectives can thus be accommodated in terms of the pragmatic dimension of understanding, in a way that
is compatible with the basic factivity requirement of (minimal) realism.

5 Implications for Scientific Realism

The case of the rainbow typifies the way in which scientific understanding
is situated in and colored by radically different ways of thinking about
what there is in the world and what laws of nature describe. Our current
science provides one set of perspectives, and we should be open to yet
different, further theoretical perspectives that may be conceived in the
fullness of time. In so far as scientific realism involves a commitment to
genuine scientific understanding and progress thereof, it must embrace
and make sense of such explanatory perspectives.

Notwithstanding the plurality of explanatory perspectives, there is a
standard story of the accumulating understanding of the rainbow due
to Descartes, Newton, Young, Fresnel, and many others. I have argued
that a well-motivated philosophical account of explanatory understanding vindicates and further explicates this story. The realist dimension
of this account is due to the assumption that genuine explanations are

underwritten by explanatory dependencies in the world. This is the basic factivity requirement of the counterfactual-dependence account. Explanatory understanding, in turn, can be construed as an agent's ability to make correct counterfactual *what-if* inferences. Thus construed, understanding has several distinctly pragmatic aspects, which can be associated with the non-factive elements of explanatory theories and models, such as idealizations and mistaken metaphysical presuppositions, that are involved in different explanatory perspectives. From the viewpoint of this account, a realist can make sense of the steady accumulation of genuine understanding of various optical phenomena, including the paradigmatic rainbow, regardless of the fact that all explanations are situated in one or another theoretical perspective.

Clearly, this realist account is rather minimal in its commitments to what the unobservable world is like. In particular, it does not incorporate the ("standard" realist) notion that our current best theories are "approximately true," or that they approach some kind of "ultimate" (God's-eye) perspective. And it neither supports nor presupposes inference to (the approximate truth of) the best explanation. (Indeed, as we have seen, what counts as "best" partly depends on the context in which explanations are given and assessed.) At the same time, the kind of understanding that we can attribute to scientists satisfies (suitably minimal) realist ambitions given the factivity assumption, and it certainly goes beyond antirealism according to which theories of light are merely effective instruments for making predictions of observable phenomena and guiding practical applications and interventions.[24] The factivity assumption requires that genuine understanding is underwritten by objective worldly facts about how the explanandum really depends on the explanans. Understanding accumulates when our explanatory theories and models give us the ability to make more *what-if* inferences, the correctness of which corresponds to worldly dependence facts. This accumulation can be partly a matter of new explanations containing explanatory information in a cognitively more salient form, given our cognitive makeup, inferential abilities, and training. And, more importantly for the realist, the accumulation of understanding is often a matter of introducing new explanatory variables that represent further explanatory dependencies, typically in the form of functional equations linking the explanans and the explanandum. These variables capture physical features of the world that need not be observable. In a given theoretical perspective, these variables can be given a rich metaphysical interpretation, which the realist should not be committed to. Or, these variables can be presented in an idealized way, which the realist should not be committed to either. Rather, she should only be committed to the most minimal interpretation of the explanatory variables that allows her to speak of the explanatory dependencies.

In the case of the rainbow, this commits the realist to explanatory variables that capture properties of light and rain, such as the shape of the raindrops, their size relative to light's wavelength, and the direction of light's propagation and polarization. The fact that there are various perspectives in which such explanatory variables have been embedded is an essential part of the way *human beings* understand the world, involving also non-factive aspects of our explanatory theories and models. Exactly which aspects are non-factive? We do not know. Some aspects of explanations are justifiably regarded as non-factive idealizations, given their discord with our theoretical beliefs and (typically) the prospects of de-idealization. But on the whole we are not reliably able to sharply demarcate between our current explanations' factive and non-factive aspects, especially when it comes to the interpretation of the variables that feature in functional relationships that support explanations in the counterfactual-dependence mode. This is a lesson we have to learn from the history of science, as some of the more minimally inclined (for example, structural) realists have acknowledged (Saatsi 2019). At the same time, nothing in the history of science speaks against the broader realist notion that steady and genuine explanatory progress is being made with understanding-providing theories and models that de facto latch on to reality better and better (in the sense of the basic factivity requirement). This progress in scientific understanding of the world does not amount to accumulating knowledge, however, since understanding is not factive in the way knowledge is, and non-factive aspects of theoretical representations can increase our understanding without us knowing their non-factive status. Thus while I agree with the self-proclaimed "perspectival realists" that a notion of "perspective" helps in articulating scientific realism, I do not think we should necessarily associate this notion with knowledge the way they do.

Acknowledgments

I would like to thank Callum Duguid, Steven French, Kareem Khalifa, Rob Knowles, Michela Massimi, and Alice Murphy for very helpful comments on an earlier draft. Support from The British Academy as part of the Mid-Career Fellowship *Scientific Realism Reinvigorated* is gratefully acknowledged.

Notes

1. See, e.g., Giere (2006), Massimi (2018a), and Teller (2018). Massimi (2018b) offers a review.
2. This account of progressive scientific understanding dovetails with my view that theoretical progress in science in general does not reduce to accumulation of knowledge. See Saatsi (2019).
3. See, e.g., Hitchcock and Woodward (2003), Woodward (2003b), Woodward and Hitchcock (2003), Ylikoski and Kuorikoski (2010), and Jansson and Saatsi (2017).

4. See, e.g., Saatsi and Pexton (2013), Reutlinger (2016), Saatsi (2018a), Jansson and Saatsi (2017), and French and Saatsi (2018).
5. Ylikoski and Kuorikoski (2010) also present "factual accuracy" as an aspect of explanatory power. In my presentation this is built into the characterization of explaining being a matter of *correctly* answering *what-if questions*.
6. See Bokulich (2016) for related, more general discussion of the usability of the Newtonian gravitational theory in oceanographical explanations, in the context of which "the classical Newtonian force picture does [the best] job of making transparent the relevant patterns of counterfactual dependence" (Bokulich 2016, 273).
7. Woodward (2003a) has called a position along these lines "instrumental realism."
8. Experiments with prisms and such have of course been central to the scientific study of the rainbow from the Middle Ages onwards.
9. Such accumulation is recognized in the standard history of this area of science, as told by both historians of science and scientists themselves. See, for example, Boyer (1959); Haußmann (2016).
10. The antisolar point is the point on the celestial sphere that is directly opposite the sun from an observer's perspective.
11. Clearly Descartes was not in a position to answer correctly any appreciable range of *what-if* questions regarding these variables; for example, what if raindrops were oblate thus and so, as opposed to being spherical? Thus, his explanation should be considered quite shallow. But for a realist it marks a genuine explanatory advance nevertheless.
12. This is indeed something that Christiaan Huygens explicitly calculated in 1652. The answer is that the bow angle would be approximately 25° instead of 41° (Boyer 1959).
13. In due course this variable comes to be associated with light's wavelength.
14. As Newton put it: "the Science of Colors becomes a speculation as truly mathematical as any other part of *Opticks*" (Boyer 1959, 241).
15. For the primary rainbow, the single internal reflection angle near the caustic is very close to Brewster's angle, at which no p-polarized light (corresponding to the radial direction as seen from the observer) is reflected.
16. See Saatsi (2015) for further discussion of Laudan's reasoning and its limitations as an argument against realism.
17. A natural raindrop typically resembles an asymmetrically squashed sphere due to air resistance (the bottom part being flatter than the dome-like top, like a hamburger bun), so the optical scattering properties for real-life rain showers differ from those of a collection of perfect spheres.
18. See note 10.
19. The generalized Lorentz-Mie theory goes beyond plane waves and spherical drops (Gouesbet and Gréhan 2011).
20. The details that I must brush over here are well summarized in Pincock (2011) and reviewed in more detail in, for example, Adam (2002) and Nussenzveig (1992). I broadly agree with Pincock's assessment of the explanatory virtues of these models, which he however expresses independently of any particular way of understanding explanation or explanatory understanding. A further important part of the scientific understanding of the rainbow, which I do not even touch here, concerns the universality of rainbow phenomena over variation in, for example, drop shapes. See Batterman (2001, 2005) and Belot (2005).
21. Of course, Maxwell's theory does not provide a truly fundamental theory of light, since it is not a quantum theory.

22. In this tune, an epigraph in Nussenzveig's classic exposition of the CAM approach reads: "it is nice to know that the computer understands the problem, but I would like to understand it too" (Nussenzveig 1992, 37).
23. The Debye approximation, upon which the CAM approach builds, is not valid for non-spherical drops.
24. Here I differ from de Regt (2017), whose account of understanding also emphasizes the contextual nature of understanding, but articulates it in a way that is empty of any realist commitment. See also de Regt and Gijsbers (2017). Unfortunately, I don't have space to engage here with de Regt's account, which I regard as insufficient for making sense of the intertheoretic relations between different theories and models of light. See also Khalifa (2017) and Woodward (2003a).

References

Adam, J. A. 2002. "The Mathematical Physics of Rainbows and Glories." *Physics Reports* 356(4–5): 229–365.

Batterman, R. W. 2001. *The Devil in the Details: Asymptotic Reasoning in Explanation, Reduction, and Emergence.* New York: Oxford University Press.

Batterman, R. W. 2005. "Response to Belot's 'Whose Devil? Which Details?'" *Philosophy of Science* 72(1): 154–163.

Belot, G. 2005. "Whose Devil? Which Details?" *Philosophy of Science* 72(1): 128–153.

Bokulich, A. 2016. "Fiction as a Vehicle for Truth: Moving Beyond the Ontic Conception." *The Monist* 99(3): 260–279.

Boyer, C. B. 1959. *The Rainbow: From Myth to Mathematics.* New York: T. Yoseloff.

Dales, R. C. 1973. *The Scientific Achievement of the Middle Ages.* Philadelphia: University of Pennsylvania Press.

de Regt, H. W. 2017. *Understanding Scientific Understanding.* New York: Oxford University Press.

de Regt, H. W., and Gijsbers, V. 2017. "How False Theories Can Yield Genuine Understanding." In *Explaining Understanding: New Perspectives From Epistemology and Philosophy of Science*, edited by Grimm, S. R., Baumberger, C., and Ammon, S., 50–75. New York: Taylor and Francis.

French, S., and Saatsi, J. 2018. "Symmetries and Explanatory Dependencies in Physics." In *Explanation Beyond Causation: Philosophical Perspectives on Non-Causal Explanations*, edited by Reutlinger, A., and Saatsi, J., 185–205. New York: Oxford University Press.

Giere, R. N. 2006. *Scientific Perspectivism.* Chicago: University of Chicago Press.

Gouesbet, G., and Gréhan, G. 2011. *Generalized Lorenz-Mie Theories.* Berlin: Springer-Verlag.

Haußmann, A. 2016. "Rainbows in Nature: Recent Advances in Observation and Theory." *European Journal of Physics* 37(6): 063001.

Hitchcock, C., and Woodward, J. 2003. "Explanatory Generalizations, Part II: Plumbing Explanatory Depth." *Noûs* 37(2): 181–199.

Jansson, L., and Saatsi, J. 2017. "Explanatory Abstractions." *The British Journal for the Philosophy of Science*, axx016.

Khalifa, K. 2017. *Understanding, Explanation, and Scientific Knowledge.* Cambridge: Cambridge University Press.

Laudan, L. 1981. "A Confutation of Convergent Realism." *Philosophy of Science* 48(1): 19–49.

Massimi, M. 2018a. "Four Kinds of Perspectival Truth." *Philosophy and Phenomenological Research* 96(2): 342–359.

Massimi, M. 2018b. "Perspectivism." In *The Routledge Handbook of Scientific Realism*, edited by Saatsi, J., 164–175. London: Routledge.

Nussenzveig, H. M. 1992. *Diffraction Effects in Semiclassical Scattering*. Cambridge: Cambridge University Press.

Pincock, C. 2011. "Mathematical Explanations of the Rainbow." *Studies in History and Philosophy of Science Part B: Studies in History and Philosophy of Modern Physics* 42(1): 13–22.

Reutlinger, A. 2016. "Is There a Monist Theory of Causal and Noncausal Explanations? The Counterfactual Theory of Scientific Explanation." *Philosophy of Science* 83(5): 733–745.

Saatsi, J. 2015. "Historical Inductions, Old and New." *Synthese*. https://doi.org/10.1007/s11229-015-0855-5.

Saatsi, J. 2018a. "On Explanations From Geometry of Motion." *The British Journal for the Philosophy of Science* 69(1): 253–273.

Saatsi, J. 2018b. "Realism and the Limits of Explanatory Reasoning." In *The Routledge Handbook of Scientific Realism*, edited by Saatsi, J., 200–211. London: Routledge.

Saatsi, J. 2019. "What Is Theoretical Progress of Science?" *Synthese* 196(2): 611–631.

Saatsi, J., and Pexton, M. 2013. "Reassessing Woodward's Account of Explanation: Regularities, Counterfactuals, and Noncausal Explanations." *Philosophy of Science* 80(5): 613–624.

Teller, P. 2018. "Referential and Perspectival Realism." *Spontaneous Generations: A Journal for the History and Philosophy of Science* 9(1): 151–164.

Woodward, J. 2003a. "Experimentation, Causal Inference, and Instrumental Realism." In *The Philosophy of Scientific Experimentation*, edited by Radder, H., 87–118. Pittsburgh: University of Pittsburgh Press.

Woodward, J. 2003b. *Making Things Happen: A Causal Theory of Explanation*. New York: Oxford University Press.

Woodward, J., and Hitchcock, C. 2003. "Explanatory Generalizations, Part I: A Counterfactual Account." *Noûs* 37(1): 1–24.

Ylikoski, P., and Kuorikoski, J. 2010. "Dissecting Explanatory Power." *Philosophical Studies* 148(2): 201–219.

5 Universality and the Problem of Inconsistent Models

Collin Rice

1 Introduction

Most attempts to justify the explanatory use of idealized models appeal to the irrelevance of the features distorted and to the accuracy of the model with respect to relevant (e.g., difference-making) features (Craver 2006; Potochnik 2017; Strevens 2008; Weisberg 2013; Woodward 2003). This approach requires that there be a single set of relevant features or difference makers for the explanandum that a model accurately represents when it explains. However, this approach conflicts with the widespread use of *multiple conflicting idealized models* to explain and understand the same phenomenon (Longino 2013; Mitchell 2009; Morrison 2011, 2015; Weisberg 2007, 2013). Indeed, the accurate representation of relevant features approach is difficult to square with the use of a plurality of models that make conflicting assumptions and purport to represent incompatible causal structures, difference makers, and ontologies. Even if we adopt the perspectivalist's idea that model M only represents system S in a particular way from the perspective of theory T, the problem of multiple inconsistent models does not go away, since multiple perspectival models will often make incompatible ontological claims about the same target system(s) (Giere 2006; Massimi 2018; Morrison 2015, chap. 5). In response, in this chapter I argue for an alternative way to think about the model-world relation that appeals to *universality* classes to justify the use of multiple conflicting models to explain and understand the same phenomenon. The term 'universality' comes from mathematical physics, but in its most general form it is just an expression of the fact that many systems that are (perhaps extremely) heterogeneous in their physical features will nonetheless display similar patterns of behavior that are largely independent of their physical details (Kadanoff 2000, 2013). The systems that display similar patterns of behavior despite differences in their physical features are said to be in the same universality class. Using examples from biology and physics, I illustrate how universality classes can link multiple conflicting (perspectival) models to their target system(s) in ways that allow for the development of explanations and understanding. In addition, I argue that this universality approach is compatible with a factive conception of scientific understanding that is based on grasping true modal information. Maintaining a factive conception of understanding allows for a version of scientific realism to be preserved despite the plurality of explanations and understanding scientists acquire from conflicting idealized models.

Section 2 of this chapter lays out the problem of inconsistent models and shows how it raises a challenge to several accounts of how to justify the use of idealized models to explain—including perspectivalism. Then section 3 presents my alternative account that appeals to universality classes to justify the use of idealized models to explain (and understand). Section 4 shows how the universality account can provide a solution to the problem of inconsistent models for perspectivalism and applies the account to cases from biology and physics. Finally, section 5 shows how the universality account allows the use of inconsistent idealized models to be compatible with a version of scientific realism based on grasping true modal information.

2 The Problem of Inconsistent Models

Idealized models are widespread in science, which raises a philosophical puzzle: how can we extract reliable information from representations we know to contain false assumptions, which often distort relevant features of the target system? An additional layer is added when we consider the fact that, in many instances, multiple idealized models with conflicting assumptions are used to study the same phenomenon (Longino 2013; Morrison 2015; Mitchell 2009; Weisberg 2007, 2013). Rather than simply modeling different features (or aspects) of the target system in complementary ways, genuinely inconsistent models often make contradictory assumptions about the target system (e.g., the nucleus), yield incompatible causal claims, and represent the system's basic ontology in fundamentally inconsistent ways (Morrison 2011, 2015). Indeed, Michael Weisberg characterizes the common practice of *multiple model idealization* as "the practice of building multiple related but incompatible models, each of which makes distinct claims about the nature and causal structure giving rise to a phenomenon" (Weisberg 2007, 645). While Weisberg focuses on the need for these different models to satisfy various competing representational goals, the most pressing challenge for the realist comes from attempts to use multiple inconsistent models *to explain and understand* the same phenomenon because both explanation and understanding are typically thought to be *factive*—that is, they are both thought to have truth or accuracy requirements.

2.1 The Problem of Inconsistent Models and Accurate Representation Requirements

Among philosophers of science, it is widely accepted that a necessary condition for something to explain is that it be, at least in some sense, true (Hempel 1965). This intuition is based on examples such as explaining the weather by citing Greek gods. Intuitively, although this would explain *if it were true*, there is a sense in which it fails to provide a satisfactory explanation of the weather. Hempel's original Deductive-Nomological (DN) account builds this in due to the fact that, in order to be sound (or cogent), an argument must have all true premises. Hempel distinguished between genuine explanations, where the explanans is true, and potential explanations, which would be adequate if they were true. In modeling terms, this truth requirement claims that in order for a model to explain, it must *accurately represent the explanatorily relevant features of the target system(s)*.[1]

Many contemporary accounts of how models explain make this truth requirement explicit. For example, mechanistic accounts of explanation typically include particularly strong accurate representation requirements in order for a mechanistic model to explain, as is illustrated by David Kaplan and Carl Craver's model-to-mechanism-mapping (3M) requirement:

(3M) A model of a target phenomenon explains that phenomenon to the extent that (a) the variables in the model correspond to identifiable components, activities, and organizational features of the target mechanism that produces, maintains, or underlies the phenomenon, and (b) the (perhaps mathematical) dependencies posited among these (perhaps mathematical) variables in the model correspond to causal relations among the components of the target mechanism.

(Kaplan 2011, 347)

The 3M requirement involves various kinds of "correspondence" between the model and the actual causal mechanisms in the target system. Kaplan adds that "3M aligns with the highly plausible assumption that the more accurate and detailed the model is for a target system or phenomenon the better it explains that phenomenon" (Kaplan 2011, 347). In general, mechanistic accounts typically require models to provide an accurate representation of the relevant components and interactions involved in the causal mechanisms that actually produced the explanandum (Craver 2006; Kaplan and Craver 2011).

Accurate representation is also built into most causal accounts of explanation. As Michael Strevens puts it, "no causal account of explanation—certainly not the kairetic account—allows non-veridical models to explain" (Strevens 2008, 297). On Strevens's account, "a standalone explanation of an event *e* is a causal model for *e* containing only difference-makers for *e*" in which, "the derivation of *e*, mirrors a part of the causal process by which *e* was produced" (Strevens 2008, 71–72). While Strevens's account does allow some idealized models to explain when they only distort causal factors that do not make a difference, accurate representation continues to play a key role, since "the overlap between an idealized model and reality . . . is a standalone set of difference-makers for the target" (Strevens 2008, 318). In addition, James Woodward suggests that causal models explain when they "correctly describe," "trace or mirror," or are "true or approximately so" with respect to the difference-making causal relations that hold between the explanans and the explanandum (Woodward 2003, 201–203). More generally, for causal accounts, in order for a model to explain, it must provide an accurate representation of the difference-making causal relationships (or causal processes) within the model's target system(s). Indeed, Michael Weisberg describes a wide range of accounts of minimalist idealization, which claim that models that explain are those that "accurately capture the core causal factors," since "[t]he key to explanation is a special set of explanatorily privileged causal factors. Minimalist idealization is what isolates these causes and thus plays a crucial role for explanation" (Weisberg 2007, 643–645).

Some causal accounts involve less demanding accurate representation requirements. For example, Angela Potochnik's (2015, 2017) recent causal account allows idealized models that explain to distort some causal difference makers. However, while Potochnik's view does allow for some causal difference makers to be left out or idealized, her account still requires models that explain to accurately represent the causal factors that had a significant impact on the causal pattern of interest to the current research program. She argues that "posits central to representing a focal causal pattern in some phenomenon must accurately represent the causal factors contributing to this pattern. . . . Idealizations, in contrast, must . . . represent as-if [such that . . .] none of its neglected features interferes dramatically with that pattern" (Potochnik 2017, 157). Therefore, while the set of causal factors is delimited somewhat differently, there is still a specific set of significant causal factors that needs to be accurately represented in order for an idealized model to explain. Moreover, similar to Strevens's view, idealizations are justified in these explanations when they distort features that are irrelevant to the causal pattern of interest.

In addition to these accurate representation requirements for models to explain, philosophers of science have long recognized a strong connection between explanation and understanding (de Regt 2009; Grimm 2008; Strevens 2013). For example, Wesley Salmon writes: "understanding results from our ability to fashion scientific explanations" (Salmon 1984, 259). An even stronger position is adopted by J. D. Trout, who claims that "scientific understanding is the state produced, and *only* produced, by grasping a true explanation" (Trout 2007, 585–586, emphasis added). Michael Strevens also argues that an individual has scientific understanding of a phenomenon only if they grasp the correct scientific explanation of that phenomenon (Strevens 2008, 2013). The link between explanation and understanding is also echoed by epistemologists: "understanding why some fact obtains . . . seems to us to be knowing propositions that state an explanation of the fact" (Conee and Feldman 2011, 316). While I think there are good reasons for believing that understanding is possible without having an explanation (Lipton 2009; Rohwer and Rice 2013, 2016), the fact that much of our scientific understanding comes from providing explanations that are thought to be accurate representations of relevant features lends support to the idea that scientific understanding is typically thought to be factive in some sense as well (at least in cases of understanding why). Indeed, like the intuition that knowledge of a proposition requires that the proposition be true, most epistemological accounts of understanding maintain that genuine understanding must be factive in some way (Grimm 2006; Khalifa 2012, 2013; Mizrahi 2012; Kvanvig 2003, 2009; Rice 2016; Strevens 2013).[2]

In general, most philosophical accounts claim that for idealized models to provide explanations and understanding, the accurate representation relation must hold between the idealized model and the important, significant, or difference-making causes (or mechanisms) that actually produced the explanandum. Given these factive requirements for explanation and understanding, many philosophers have noted that the use of multiple inconsistent models for the same phenomenon raises serious challenges for scientific realism. Indeed, if the above accounts are right, then when multiple conflicting models are used to explain and understand the same phenomenon, they ought to be interpreted as making conflicting claims about the causal interactions among the relevant features of the models' target system(s). In other words, multiple conflicting models that are used to explain and understand the same phenomenon ought to be interpreted as each aiming to provide an accurate representation of the difference-making causes (or mechanisms) for the explanandum. This is problematic given that the idealized models used in science often make conflicting assumptions (Morrison 2011, 2015), represent incompatible causal structures (Longino 2013), and distort difference-making causes in a variety of ways (Batterman and Rice 2014; Rice 2018, 2019). The problem of inconsistent models, then, is to show how such a situation can result in genuine explanations and understanding of the target phenomenon, despite the inconsistency of the representations appealed to by scientific modelers. In light of the above discussion, I argue that the problem of inconsistent models has largely resulted from the fact that most accounts of how to justify the use of idealized models to explain and understand are exclusively focused on *accurate representation of relevant features* of the target system.

2.2 *The Problem of Inconsistent Models and Perspectivalism*

One possible response to the challenge of inconsistent models comes from perspectivalism (Giere 2006; Massimi 2018). Giere's original perspectivalism suggests that we should not directly interpret models as making claims about their target systems. Instead, we should interpret models as contextually constructed representations of a system from the perspective of a particular theory. Models constructed within different theoretical perspectives will thus make different and sometimes contradictory claims about the same target system(s). In short, perspectivalism suggests that we need to recast claims about how models represent in the following form: from the perspective of theory T, model M represents system S in a particular way (Giere 2006).

While this perspectival move is somewhat helpful in showing how multiple models might represent the system in different context-sensitive ways, I agree with Morrison (2011, 2015) and Massimi (2018) that perspectivalism on its own does little to solve the problem of inconsistent models for realism. For one thing, even if we can only interpret how (or what) models represent from within the perspective of a particular theory or context, this doesn't tell us how sets of inconsistent models from different perspectives are able to yield genuine explanations and understanding of the same phenomenon. Several epistemological and methodological questions remain about how to interpret the inconsistent claims made by multiple perspectival models. But more than just failing to provide a clear solution, I contend that perspectivalism (as well as appeals to partial structures or structural realism) will continue to fall prey to the problem of inconsistent models as long as philosophical accounts of modeling continue to conflate the following two questions:

1. How does the idealized model allow scientists to *explain and understand* the phenomenon?
2. Which of the relevant features (e.g., difference-making causes) for the occurrence of the phenomenon are *accurately represented* by the idealized model?

In short, the problem lies with equating explanation and understanding with accurate representation of relevant features, due to the truth and accuracy requirements maintained by most accounts of explanation and understanding outlined above. As long as these accurate representation requirements are central to how philosophers conceive of explanation and understanding, appealing to only a limited set of difference makers (Potochnik 2017), partially accurate representations (Wimsatt 2007; Worrall 1989), or perspectivalism (Giere 2006) will not be able to offer a solution to the problem of inconsistent models for realism. After all, if the models are still intended to accurately represent (or map on to) relevant features of the target system, multiple conflicting idealized models will lead to inconsistent metaphysical claims about the components and inter-actions of the target system.

In support of this suggestion, Michela Massimi (2018) has shown how the main argument used to conclude that the use of multiple inconsistent perspectival models is incompatible with realism depends on what she calls the "representationalist assumption," namely that scientific models (partially) represent relevant aspects of a given target system. Further-more, this representationalist assumption implicitly depends on the idea that "representation means to establish a one-to-one mapping between relevant (partial) features of the model and relevant (partial)—actual or fictional—states of affairs about the target system" (Massimi 2018, 342). This kind of accurate representation (or mapping) assumption results in inconsistent models making inconsistent metaphysical claims about the target system even if we allow that models represent the target system from different perspectives. In short, the reason perspectival modeling is similarly challenged by the problem of inconsistent models is that it is typically taken to include the same kinds of assumptions regarding accurate representation of (or mapping on to) relevant features of the target system. This again shows that the problem of inconsistent models derives largely from the assumption that models that are used to explain and understand a phenomenon must accurately represent the relevant features of the target system.[3]

As additional motivation for moving away from accurate representa-tion assumptions, Morrison notes that, in cases of inconsistent models, "we usually have no way to determine which of the many contradictory models is the more faithful representation, especially if each is able to generate accurate predictions for a certain class of phenomena" (Morrison 2011, 343). Therefore, not only is accurate representation the source of the challenge from inconsistent models, but in many cases we simply can-not answer the question of which of the plethora of idealized models accu-rately represents the relevant features of the models' target system(s). As a result, I contend that we ought to consider alternative ways of justifying the use of multiple inconsistent models to explain and understand the same phenomenon that do not depend on the accurate representation relations that are central to most accounts of how models provide expla-nations and understanding.

3 Using Universality Classes to Justify the Use of Idealized Models

Instead of focusing on accurate representation relations, I argue that a better way to justify the use of (multiple conflicting) idealized models to explain and understand is to appeal to *universality classes* (Batter-man 2000, 2002; Batterman and Rice 2014; Rice 2018, 2019). The term 'universality' comes from mathematical physics, but in its most general form it is just an expression of the fact that many systems that are perhaps extremely heterogeneous in their physical features will nonetheless display similar patterns of behavior (typically at some macroscale) (Batterman 2000, 2002; Kadanoff 2013). The systems that display similar behaviors despite differences in their physical features are said to be in the same universality class (Kadanoff 2013). While most instances of universality have focused on patterns that are stable across real systems— for example, the universality of critical exponents across a wide range of fluids—a more general conception of universality can be used to find classes of real, possible, *and model* systems that display similar behaviors despite differences in their features.[4] I contend that this link of being in the same universality class can justify the use of idealized models to explain and understand the behaviors of their target systems even when the models fail to accurately represent the relevant features of those systems.

Given that many idealized model systems are in the same universality class as their target systems, they will display similar patterns of behavior despite the fact that the model may drastically distort the causes, mechanisms, or other features responsible for the explanandum in any real-world system. Indeed, just like cases of universality that show stability across perturbations of the features of extremely different real systems (e.g., fluids and magnets), universality can also link stable behaviors across perturbations of the features of different models systems and their real (or possible) target systems. According to what I will call the 'universality account,' it is precisely this stability of various behaviors across perturbations of most of the system's features that can enable scientists to justifiably use idealized models that drastically distort difference-making features to explain and understand the behaviors of their target system(s) (Batterman and Rice 2014; Rice 2018, 2019).

There are several important things to note about this universality account. First, idealized models that accurately represent difference-making features can clearly be justifiably used to explain and understand on the universality account. After all, if those difference-making features are sufficient to produce the phenomenon in some real-world systems, then those features will likely be sufficient for producing the phenomenon in the model system as well (although a few additional assumptions may be required). Thus, those models will be in the same universality class as their target system(s). Consequently, the universality account can easily accommodate the cases used to motive accounts that focus on accurate representation relations. Importantly, however, the universality account shows that accurate representation is not what is doing the important justificatory work in using these idealized models to explain and understand—rather it is the fact that the idealized model produces the universal patterns of behavior that are of interest to the modeler and are stable across a class of real, possible, and model systems. Indeed, the universality account allows scientists to justifiably use idealized models to explain and understand even in cases where the same universal patterns of behavior are produced by extremely *different* sets of features across the real, possible, and model systems within the universality class. This allows the account to justify using idealized models to explain and understand a phenomenon even when we cannot identify which parts of the model are accurate and which parts are distorted representations of the target system. This is quite useful since in many cases the accurate parts of the model cannot be isolated from the idealizing assumptions used in constructing the model; that is, it is often difficult (or impossible) to decompose scientific models into their accurate and inaccurate parts (Rice 2019). The universality account also provides a way around the problem posed by Morrison, that "we usually have no way to determine which of the many contradictory models is the more faithful representation" (Morrison 2011, 343). Since idealized models within the same universality class as their target system(s) need not accurately represent the features of their target system, scientists can still be justified in using those models to explain and understand even when they are unable to determine which model is the most accurate or which features are being accurately represented by individual models. In short, universality classes provide a way of linking the behaviors of idealized models with the behaviors of their target systems without relying on the accurate representation relations required by most accounts of how idealized models can be used to explain and understand.

3.1 Modal Information, Explanation, and Understanding

Before going further, it is important to say a bit about why the information scientists obtain from idealized models within a universality class can be used to explain and understand the behaviors of real systems. Indeed, simply "reproducing a pattern of behavior" might sound like these are merely phenomenological models that produce the desired results but fail to explain why the phenomenon occurs. In what follows, I focus on the explanations and understanding produced by *providing true modal information about the phenomenon*. I argue that idealized models can be used to provide explanations and understanding because they not only produce the behaviors of interest but also enable scientists to identify which features are important for the occurrence of the explanandum, and how various changes in those relevant (and irrelevant) features would result in changes in the explanandum (or not). That is, these idealized models provide extensive modal information about the phenomenon of interest concerning counterfactual dependencies and independencies that hold in the model's target system(s). For example, an optimization model that is within the same universality class as a real population can show biologists how changes in the tradeoffs between various fitness-enhancing features within a system will result in changes in the expected equilibrium outcome—even if the model represents those features and the relationships between them and the explanandum in a highly idealized (i.e., distorted) way (Rice 2013). Moreover, by showing that a class of real, possible, and model systems all display similar patterns of behaviors despite differences in their features, scientists can come to understand that many of the features of the system are counterfactually irrelevant to the explanandum. For example, by investigating an optimization model a biologist can determine that the equilibrium point of the population is independent of the starting point, trajectory, or method of inheritance in the population (Potochnik 2007, 2017; Rice 2013, 2016).

Accordingly, on the account I present here, idealized models within the same universality class as their target system(s) can be used to provide explanations by revealing a plethora of modal information about the counterfactual dependencies (and independencies) between features of real system(s) and the explanandum.[5] My appeal to modal information is not meant to provide a complete or exhaustive account of explanation—that is, there are certainly additional criteria required of explanations, and there may be explanations that do not involve modal information—but it is worth noting that modal information is central to many causal and non-causal accounts of explanation (Batterman and Rice 2014; Bokulich 2011, 2012; Kim 1994; Rice 2013; Woodward 2003). Indeed, many accounts of explanation agree with Woodward that "[an] explanation must enable us to see what sort of difference it would have made for the explanandum if the factors cited in the explanans had been different in various possible ways" (Woodward 2003, 11). That is, explanations are provided by giving counterfactual (i.e., modal) information about how changes in certain features in the explanans result in changes in the explanandum. This is precisely the kind of information that scientists are able to extract from model systems within the same universality class as their target system(s). The key feature of models that can be used to develop explanations is that they provide a set of modal information that captures how various changes in the explanatorily relevant features of the target system would result in changes in the explanandum.

In addition, I argue that *understanding* is also provided by grasping modal information. For example, understanding can be produced by idealized models that investigate other (perhaps very distant) scenarios in the network of possibilities, that is, the range of possible states of the system. By providing modal information about possible systems, models that fail to explain may still be able to produce understanding of a phenomenon (Lipton 2009; Rohwer and Rice 2013; Rice 2016). This approach builds on a proposal by Robert Nozick:

> I am tempted to say that explanation locates something in actuality, showing its actual connections with other things, while *understanding*

locates it in a network of possibility showing the connections it would have to other nonfactual things or processes. (Explanation increases understanding too, since the actual connections it exhibits are also possible.)

(Nozick 1981, 12)

Nozick's suggestion nicely links the understanding produced by explanations with the understanding produced by models that fail to explain (Lipton 2009; Rice 2016). In both cases, understanding is produced by providing true modal information about the phenomenon of interest—in Nozick's terminology, it "locates it in a network of possibility." However, I argue that in the case of understanding produced by an explanation, a particular and more expansive set of modal information about the explanandum must be provided that identifies a set of relevant features responsible for the explanandum and how changes in those features would result in changes in the explanandum (Bokulich 2011, 2012; Rice 2013; Woodward 2003). This modal information improves understanding by telling scientists about the range of possible states of the system that would (and would not) produce the explanandum. Still, modal information not included in an explanation can also improve understanding of the possible states of the system(s) of interest, for example, by identifying which features are not necessary for the phenomenon to occur (Rohwer and Rice 2013).

What is crucial to notice is that the modal information involved in explanation and understanding can be provided in ways *other than accurately representing relevant features*. An idealized model can tell a scientist quite a lot about how things would be in various counterfactual scenarios without having to accurately represent the features of any actual system(s). In other words, idealized models can provide modal information about changes in features of the system even when they fail to accurately represent those features or the actual processes that link those features to the target explanandum. I argue that in many cases this is possible because universality guarantees that the model system's patterns of counterfactual dependence and independence will be similar to those of the target system(s), even if the entities, causes, and processes of those systems are extremely different. Therefore, even if the model drastically distorts the relevant features of its target system(s), it can still be used to explain or understand the phenomenon because many of the modal patterns that hold in the idealized model system will be similar to those of the real-world system(s). For example, in the case of critical behaviors in physics, investigation of the universality class reveals that the explanandum counterfactually depends on the dimensionality of the system and the symmetry of the order parameter (Batterman forthcoming). Furthermore, by using renormalization techniques to explicitly delimit the universality class, scientists can demonstrate that most of the other physical features of the systems in the class are counterfactually irrelevant to the explanandum (Batterman 2002). Generalizing the concept of universality allows us to capture this stability of various patterns of behavior that are largely independent of the physical components, interactions, and features of a heterogeneous class of real, possible, and model systems. Moreover, by focusing on modal information we can see how highly idealized models (and various modeling techniques) can allow scientists to extract a plethora of modal information that can be used to explain and understand a phenomenon.

In summary, suppose that we could show that:

1. Certain (modal) patterns of behavior are universal across classes of real, possible, and model systems.
2. The idealized models used by scientists are within the same universality classes as the real systems whose behaviors they want to explain and understand.

If this were the case, we could provide an epistemic justification for why those idealized models can be used to generate explanations and understanding of phenomena in real-world systems, despite providing drastically distorted representations of their target system(s).[6] Importantly, this justification does not require the accurate representation relations involved in most other accounts of how to justify the use of idealized models to explain and understand.

4 Universality and Multiple Conflicting Models

Having now presented the universality account, it is time to return to the problem of inconsistent models. While I do not intend to present the universality account as a univocal account of how models relate to their target systems (I suggest a more pluralistic approach to model-world relations), I do argue that one reason to favor the universality account is that it provides a way of seeing how conflicting sets of idealized models can be used to explain and understand the same phenomenon. The solution appeals to *overlapping universality classes* that each include the target system(s) in which the phenomenon occurs. Sets of universality classes can link sets of inconsistent models with the same real-world phenomenon without having to have each model within the same universality class (Figure 5.1).

Moreover, since universality classes link models with their target systems without requiring accurate representation of (or mapping onto) relevant features, this account need not result in the inconsistent metaphysical claims that plague other accounts of how idealized models connect with their target systems. Specifically, universality classes show that certain counterfactual dependencies and independencies will be stable across a class of real, possible, and model systems that are perhaps drastically

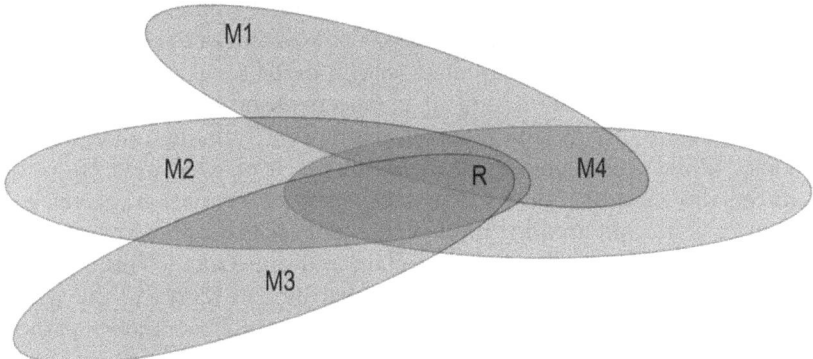

Figure 5.1 Multiple conflicting models (M1, M2, M3, M4, . . .) might be connected to the same real-world phenomena via multiple overlapping universality classes (represented by the ellipses) to which the real-world system (R) belongs. M1 and M4 show that multiple model systems might be within the same universality class as well.

heterogeneous in their features. This allows sets of models to drastically distort the features of the target system in inconsistent ways and still enables scientists to use those models to extract the modal information required to explain and understand various phenomena. Moreover, since every target system will involve a myriad of counterfactual dependencies and independencies that will be stable *across different classes of systems*, models in different universality classes can capture different sets of modal information about the same phenomenon. That is, because a single target system will be a member of multiple universality classes, models in different universality classes can connect with the target system in different ways and capture different sets of modal information about the phenomenon of interest. Perhaps one of these models will provide sufficient modal information to explain or understand the phenomenon on its own, or perhaps the modal information provided by multiple models will have to be combined by scientific modelers in order to formulate the desired explanations and understanding. The key point is that claiming that an idealized model provides some true modal information about the target system is importantly different from claiming that the model provides an accurate representation of any of the system's features. Consequently, because the universality account does not depend on accurate representation of the relevant features that produced the explanandum, the use of multiple inconsistent idealized models to explain and understand the same phenomenon need not result in the kinds of metaphysical inconsistencies that trouble scientific realists.

It is also worth seeing how this solution to the problem of inconsistent models helps perspectivalism. As I argued in section 2, the source of the problem of inconsistent models—for perspectivalists and others—is the assumption that models are intended to accurately represent the relevant features of their target system. Indeed, as Massimi's (2018) deconstruction of the argument based on inconsistent models makes clear, the argument against the combination of perspectivalism and realism depends on this kind of "representationalist assumption" (among other assumptions). While representation is crucial to much of what models do, the universality account shows why accurate representation of relevant features is not required for models to enable scientists to extract modal information that can be used to explain and understand. This allows us to maintain the perspectivalist's claim that models represent the system in a particular way from the perspective of a particular theory without also claiming that such a representation needs to be characterized as an attempt to accurately represent the relevant features of the system. In this way, multiple conflicting models constructed within the perspectives of different theories can be used to provide different sets of modal information involved in scientific explanations and understanding without producing the kinds of metaphysical inconsistencies that result from trying to interpret perspectival models as intended accurate representations. In the rest of this section, I further illustrate how the universality account provides a solution to the problem of inconsistent models by looking at cases from biology and physics.

4.1 An Example From Biology

The use of multiple conflicting models to study the same phenomenon is widespread in biology, since biological modelers often require various kinds of idealized models to describe interactions at different spatial scales (Green and Batterman 2017; Qu, Garfinkel, Weiss, and Nivala 2011). In addition, biological modelers often face the challenge of modeling processes that operate on very different timescales: seconds, days, or years (Davidson, von Dassow, and Zhou 2009). For example, cellular phenomena take place across a range of scales and "each scale of cell biology not only has its characteristic types of data, but also typical modeling and simulation approaches associated with it" (Meier-Schellersheim, Fraser, and Klaushcen 2009, 4). That is, biologists investigating cellular behaviors often require multiple models due to the wide gaps between the spatial and timescales of various aspects of the target system(s), such as between the timescales of intra-molecular dynamics (10^{-2} seconds) and chemical aspects of the interactions (10^3 seconds). As a result, multiple inconsistent modeling techniques are used at different scales to investigate the same biological phenomenon. Moreover, because many biological phenomena take place *across* multiple scales, biologists often construct a variety of conflicting models for the various possible interactions across multiple scales (Green and Batterman 2017; Meier-Schellersheim et al.

2009). That is, in addition to multiple conflicting models at different scales, biologists often build multiple conflicting multiscale models as well (Dallon 2010).

These sets of models often make conflicting assumptions about the fundamental ontological components and interactions of the target system. For example, "while cellular automata models treat the single 'cells' in their simulations as entities with fixed shape and size, Potts model simulations aim at reproducing the shape changes cells undergo due to mechanical contact with neighbor cells or extracellular matrices" (Meier-Schellersheim et al. 2009, 6). Furthermore, these sets of models represent the interactions of the target system in contradictory ways, for example, modeling the same system dynamics using both individual-based and continuum models, which make contradictory assumptions about which aspects (and scales) of the system are relevant (Byrne and Drasdo 2009). Moreover, some of these models aim to represent dynamical interactions of the system with well-defined functions, while others investigate the same behaviors by simply assigning (idealized) computational algorithms to the elements of the system and studying the emerging behaviors (Qu et al. 2011). These models not only represent the fundamental nature of the entities in drastically different ways, but they also make very different assumptions about which features of the system are relevant and irrelevant to the phenomenon. Indeed, when it comes to multiscale modeling approaches in biology, "despite the attractiveness of this method, it faces many challenges, such as the gaps between models of different scales and inconsistencies between different methodologies" (Qu et al. 2011, 23).

The challenge is to figure out how to use the insights provided by these conflicting modeling approaches in order to develop explanations and understanding of the same target phenomenon. One approach is to try and construct "master models" that integrate the data from multiple scales into a single multiscale model (Meier-Schellersheim et al. 2009). However, while integrating input data from multiple scales into a single model might produce additional insights about the phenomenon, building a master model does not resolve the contradictions among the various assumptions of the multiple models used to study the phenomenon, nor does it guarantee that the master model will provide an explanation.

Instead of focusing on the construction of master models, I argue that a more promising response to these challenges is to consider how these various conflicting models can be related to the same target phenomenon by multiple overlapping universality classes. That is, these different models can belong to different (overlapping) universality classes that each contains the target system(s) of interest to the biological modelers. For example, model M_1 might be in universality class U_1 that contains the target system, while model M_2 might be in universality class U_2 that also contains the target system (see Figure 5.1 again). Some of these idealized models might also be in the same universality class as other models used to study the phenomenon, but the important point is that they can each be in at least one universality class that also contains the target system(s).[7] Because these models are in the same universality classes as the target system(s) in which the phenomenon of interest occurs, they will display some similar patterns of behavior. In addition, since different universality classes will focus on different universal patterns, models in different universality classes can be used to extract different sets of modal information about the biological phenomenon. Importantly, this account does not require that the conflicting models be interpreted as accurately representing the difference-making (or otherwise relevant) features of the target system. Moreover, because the modal information extracted from models within different universality classes need not conflict in the ways that the representations of the models themselves conflict, scientific modelers can use this plethora of modal information to construct (consistent) explanations and understanding of the phenomenon without having to build a "master model" intended to accurately represent all of the relevant features across a variety of scales.

In sum, every biological system will be a member of many overlapping universality classes. Some of these universality classes will involve universal behaviors that hold only at particular scales, some of them will hold across multiple scales, and *many will include the model systems represented by idealized biological models*. Consequently, an idealized biological model can provide modal information about those target system(s) within the same universality class even if the model conflicts with other idealized models used to study the same phenomenon (in other universality classes). Furthermore, since universality classes do not link their systems through accurate representation or mapping, we need not interpret multiple conflicting models as making conflicting metaphysical claims about the features of real systems. As a result, multiple conflicting idealized models in different universality classes can provide different sets of modal information about the target system that can be used to construct various explanations and understanding of the phenomenon of interest.[8]

4.2 An Example From Physics

A second example of using multiple inconsistent models to study the same phenomenon comes from physicists' modeling of the nucleus (Morrison 2011). There are over 30 different nuclear models, each of which provides some insight into some aspects of nuclear structure and dynamics. However, the set of assumptions made by any one of these models is in conflict with fundamental claims made by the others (Morrison 2011, 347). For example, "some models assume that nucleons move approximately independently in the nucleus . . . while others characterize the nucleons as strongly coupled due to their strong short range interactions" (Morrison 2011, 347). Indeed, widely used models for studying the nucleus, such as the liquid-drop model and the shell model, make contradictory assumptions about the fundamental nature of elements and interactions involved (Morrison 2011, 349). In addition, these conflicting assumptions are typically necessary for the models to produce the behaviors of interest to physicists—that is, without these assumptions the models would be unable to provide insights into how the nucleus gives rise to the range of observations physicists want to explain and understand. In particular, "nuclear spin, size, binding energy, fission and several other properties of stable nuclei are all accounted for using models that describe one and the same entity (the nucleus) in different and contradictory ways" (Morrison 2011, 349). As before, the problem of inconsistent models is seeing how we can interpret these sets of conflicting idealized models as providing genuine explanations and understanding of the same phenomenon.

The perspectivalist might try to interpret these inconsistent models as each only representing the nucleus in different ways from the perspective of different theories. That is, the liquid-drop model and the shell model simply represent the nucleus in different ways from within different theories, such as classical physics or quantum mechanics. However, as Morrison points out, this really isn't much of a solution to the problem of inconsistent models, since "none of those 'perspectives' can be claimed to 'represent' the nucleus in even a quasi-realistic way since they all contradict each other on fundamental assumptions about dynamics and structure" (Morrison 2011, 350). In other words, it is difficult to see how we could make the realist inference from predictive success of these models to the accuracy of the models, given that their basic assumptions conflict with one another and those assumptions are essential to the limited predictive successes of each model.

As before, I suggest the main reason this use of inconsistent models appears problematic for the realist is the (mistaken) assumption that accurate representation is essential to interpreting the models as providing explanations and understanding. It is clear that none of these nuclear models ought to be interpreted as providing an accurate representation of all the relevant (e.g., difference-making) features for nuclear phenomena. However, the universality account shows us how explanation and understanding might be achieved without providing accurate representations of the relevant features of the system. Accordingly, I suggest that we should instead interpret these various nuclear models as relating to their target system(s) via different (and sometimes overlapping) universality classes that capture different universal nuclear behaviors across a range of perturbations to the physical features of the system. For example, the liquid-drop model might be within universality class U_1 that includes real nuclei and displays a certain range of universal behaviors, whereas the shell model might be in universality class U_2 that also includes real nuclei but displays a different range of universal behaviors, and so on for the other models. Because these models are in the same universality classes as real systems that display nuclear phenomena, they will display some similar patterns of behavior as those real systems. In addition, since different universality classes will capture different universal patterns across different classes of systems, models in different universality classes can be used to extract different sets of modal information about the nucleus without having to interpret any one of the models as an accurate representation of a single set of relevant or difference-making features for the phenomenon of interest. Different models can then provide a plethora of modal information about nuclear phenomena even if no single model can provide enough information to explain and understand all of those phenomena. Indeed, while none of the models provides an accurate representation, Morrison does grant that these conflicting nuclear models have "generated information about nuclear phenomena that can be used in practical contexts" (Morrison 2011, 350). Moreover, while no single model is able to explain all the various features of nuclear phenomena, some models have provided "an explanatory foundation for understanding certain processes" (Morrison 2011, 350). I argue that we can understand how this can be despite the inaccuracy and inconsistency of the models by appealing to different universality classes for the idealized models that allow them to provide different sets of modal information about the target phenomena. Since universality classes do not link systems via accurate representation or mapping, we need not interpret these multiple conflicting models as making conflicting metaphysical claims about the features of real systems. Instead, multiple conflicting idealized models in different universality classes can provide different sets of modal information about the target system that can be used to construct various explanations and understanding.

5 Multiple Conflicting Models, Modal Information, and Realism

One question remains: how is the universality account compatible with (a perspectival version of) scientific realism, if it grants that our models and theories are typically inaccurate representations of their target systems? The key is noting that the explanations and understanding provided by multiple idealized models can be separated from the assumptions of the models themselves. According to the universality account, explanations and understanding are provided by the modal information *extracted from* idealized models within the same universality classes as their target systems. However, the modal information included in scientists' explanations and understanding of a phenomenon need not include the inconsistent assumptions included in the conflicting idealized models used to study that phenomenon. After all, the universal modal patterns of behavior are those that are stable across perturbations in many of the features of the systems within the universality class. Consequently, discovering universality classes can provide confidence that the same patterns of counterfactual dependence and independence that occur in the idealized models will be realized in the model's target system(s), even if the model's assumptions drastically distort the difference-making features of the system and conflict with the assumptions of other models used to study the same phenomenon. In short, the counterfactual dependencies and independencies that are stable across the universality class and are used to explain and understand can be separated from the detailed assumptions used to construct the model's representation of the idealized model system (Rohwer and Rice 2016).

Distinguishing the understanding provided by models from the assumptions of the models themselves allows us to maintain a factive conception of scientific understanding despite the use of idealized models in science. In contrast with this kind of view, Catherine Elgin has recently argued that a non-factive conception of understanding is required to accommodate the epistemic successes of science (Elgin 2007, 2017). Elgin argues that any kind of *veritism* that takes truth to be necessary for epistemic success is unacceptable, since "if we accept it, we cannot do justice to the epistemic achievements of science" (Elgin 2017, 9). In particular, Elgin argues that "the more serious problem comes with the laws, models, and idealizations that are acknowledged not to be true but that are nonetheless critical to, indeed at least particularly constitutive of, the understanding that science delivers" (Elgin 2017, 14). While I am sympathetic with many aspects of Elgin's views, I disagree with the claim that the understanding produced by scientific inquiry must be *partially constituted* by the idealizations used in science. Although theories, models, and idealizations are certainly the tools with which scientists produce explanations and understanding of various phenomena, it does not directly follow that the assumptions involved in those tools must be included in the explanations and understanding *extracted from* scientists' uses of those tools. If this separation between the representations used by scientists and the understanding provided by scientific inquiry is possible, then recognizing the role of (multiple conflicting) idealized models in science need not force us to adopt a non-factive conception of scientific understanding. Indeed, the universality account maintains the requirement that the modal information used in explaining and understanding a phenomenon *must be true of the phenomenon* without requiring that the models be accurate representations of their target system(s). While the details of what makes scientific understanding factive will have to be sorted out elsewhere, the universality account enables us to see how scientific models can yield lots of true modal information about a phenomenon without having to interpret the assumptions of the model as aiming to provide an accurate representation of the target system(s).

Consequently, an important implication of the universality account is that it shows how the use of multiple conflicting idealized models to study the same phenomenon can be consistent with a more nuanced form of scientific realism that focuses on grasping true modal information (Rice 2016). Specifically, this approach is compatible with a version of realism that focuses on science's ability to provide factive understanding of patterns of counterfactual dependence (and independence) rather than on the truth or accuracy of our best theories and models themselves. Despite their drastic distortion of the features of real systems, scientific models within universality classes can provide a wide range of true modal information about the counterfactual relevance and irrelevance of various features of real systems (Batterman and Rice 2014; Massimi 2018; Rice 2013, 2016, 2019). Moreover, by building multiple models that each provide different modal information about the systems in their respective universality classes, scientists can extract myriad modal information that can be used to explain and understand the phenomenon (in a variety of ways). As a result, realists can claim that science is able to achieve the epistemic successes of explanation and factive understanding despite the fact that scientific models are typically highly idealized and conflict with one another (Potochnik 2017; Rice 2016). What is more, a perspectival view of modeling can also adopt this version of scientific realism. The key is to focus on the modal information that can be extracted from multiple perspectival models within different universality classes rather than focusing on the accuracy of the models themselves. In doing so, we can see why models from different perspectives (and in different universality classes) can provide a wide range of true modal information about real systems, even if the models themselves are inconsistent and highly idealized.

6 Conclusion

I have argued that the problem of inconsistent models is a consequence of the assumption made by most accounts that, in order for an idealized model to provide an explanation or understanding, it must accurately represent the relevant features of its target system. Discovering that this is the root of the problem motivates the exploration of alternative ways that models can relate to their target systems that enable for the development of explanations and understanding. In response, I have argued that an account based on universality classes (and modal information) can avoid this kind of accurate representation assumption and the problem of inconsistent models that it produces. Finally, I have argued that this universality account is compatible with a form of realism because the modal information extracted from idealized models can be distinguished from the assumptions of the models themselves. Going forward, I suggest that perspectivalism and other accounts of modeling continue to investigate how scientists use idealized models within particular universality classes to extract the modal information required to explain and understand real-world phenomena.

Notes

1. Of course, every model will be accurate with respect to certain features and inaccurate with respect to others. However, according to most accounts, the partial representation of a model may still be an explanation if the model accurately represents the explanatorily relevant features of the target system.
2. A non-factive account of understanding has been offered by Elgin (2017), which I address in section 5.
3. I will not be arguing for perspectivalism here. However, it is worth noting that a solution to the problem of inconsistent models that moves us away from focusing on accurate representation of relevant features (e.g., difference makers) would also help to solve the challenge to perspectivalism.
4. I refer to a model system as the abstract system represented by a scientific model that includes all and only the features specified by the model (within a particular modeling context). In a sense then, model systems are just possible systems that are picked out by scientific models.
5. It is important to distinguish using universality classes to justify the use of an idealized model to explain and understand (Rice 2018, 2019) from providing a specific type of explanation that some authors have called a "minimal model explanation" (Batterman and Rice 2014). A minimal model explanation appeals to a minimal model within a universality class, *and* the explanation relies heavily on providing a detailed backstory that shows that most of the features that distinguish the systems within the universality class are irrelevant to their universal behaviors. That is, minimal model explanations focus on demonstrating the irrelevance of most of the features of the system and then use the minimal features within a highly idealized model to show that certain minimal features are necessary for the explanandum to occur. However, not all model explanations that appeal to universality classes have this structure. Moreover, contrary to Marc Lange's (2014) objections, according to Batterman and Rice (2014) and the view defended here, being in the same universality class is not sufficient for a model to provide an explanation. Being in the same universality class merely justifies appealing to an idealized model within the explanation. More is required to extract the modal information required to provide an explanation.
6. Another objection raised by Lange (2014) to Batterman and Rice's account of minimal model explanations argues that "if our demonstrating that the model and the target system are in the same universality class were sufficient to allow us to use one to explain the other, then we might just as well use the target system to explain why the model exhibits the given behavior" (Lange 2014, 296). Lange then suggests that other accounts have an easier time accounting for this explanatory asymmetry because, "unlike the target system, the minimal model involves no other features that might make for added complications" (Lange 2014, 298). However, it is unclear why this move is not available to the minimal model explanations account as well. Indeed, it seems very close to our appeals to "computational ease" as justification for using the minimal model to explain, but Lange explicitly rejects this kind of response. I fail to see why it is satisfactory to appeal to the lack of features that would make for added complications but unsatisfactory to appeal to computational ease. In addition, Lange's claims that we cannot simply reject this asymmetry because "scientific practice does not include cases where the macrobehavior of some austere, minimal model is explained partly by the macrobehavior of some messy, real-world system" (Lange 2014, 296–297). This claim about scientific practice is certainly correct, but there are plenty of explanations for this fact that do not require that *in principle*

it could never happen that way. First, scientists simply do not have access to all the features of the real systems of interest (that is why they need the model), but they do have access to the minimal model. Moreover, as I noted above, I think practical reasons regarding computational ease *are* sufficient to explain why scientists use models to explain the behaviors of real systems in practice even if they do not rule out explanations that appeal to real systems to explain the behaviors of models in principle. So Lange's objections here seem to miss their mark. Finally, it is again important to distinguish between providing a minimal model explanation and only appealing to a universality class to justify the use of an idealized model to explain. Those processes are related, but importantly different.

7. While these universality classes may overlap in that they contain the same systems, they need not overlap and can be distinguished by having different universal behaviors. For example, in physics different universality classes are identified by different critical exponents that govern the universal behaviors of the systems in the classes. See Chen, Toner, and Lee (2015), for example.

8. Multiple (conflicting) models might also be within the same universality class, but those models will typically provide similar sets of modal information about the target phenomenon since the universal behaviors they display (and the features those behaviors are independent of) will be the same across a universality class.

References

Batterman, R. W. 2000. "Multiple Realizability and Universality." *The British Journal of Philosophy of Science* 51(1): 115–145.

Batterman, R. W. 2002. *The Devil in the Details: Asymptotic Reasoning in Explanation, Reduction, and Emergence*. Oxford: Oxford University Press.

Batterman, R. W. forthcoming. "Universality and RG Explanations." *Perspectives in Science*. http://philsci-archive.pitt.edu/13460/

Batterman, R. W., and Rice, C. 2014. "Minimal Model Explanations." *Philosophy of Science* 81(3): 349–376.

Bokulich, A. 2011. "How Scientific Models Can Explain." *Synthese* 180(1): 33–45.

Bokulich, A. 2012. "Distinguishing Explanatory From Nonexplanatory Fictions." *Philosophy of Science* 79(5): 725–737.

Byrne, H., and Drasdo, D. 2009. "Individual-Based and Continuum Models of Growing Cell Populations: A Comparison." *Journal of Mathematical Biology* 58: 657–687.

Chen, L., Toner, J., and Lee, C. F. 2015. "Critical Phenomenon of the Order-Disorder Transition in Incompressible Active Fluids." *New Journal of Physics* 17(042002): 1–10.

Conee, E., and Feldman, R. 2011. "Replies." In *Evidentialism and Its Discontents*, edited by Dougherty, T., 283–323. Oxford: Oxford University Press.

Craver, C. 2006. "When Mechanistic Models Explain." *Synthese* 153(3): 355–376.

Dallon, J. C. 2010. "Multiscale Modeling of Cellular Systems in Biology." *Current Opinion in Colloid and Interface Science* 15(1–2): 24–31.

Davidson, L., von Dassow, M., and Zhou, J. 2009. "Multi-Scale Mechanics From Molecules to Morphogenesis." *The International Journal of Biochemistry & Cell Biology* 41(11): 2147–2162.

de Regt, H. W. 2009. "Understanding and Scientific Explanation." In *Scientific Understanding: Philosophical Perspectives*, edited by de Regt, H. W., Leonelli, S., and Eigner, K., 21–42. Pittsburgh: University of Pittsburgh Press.

Elgin, C. Z. 2007. "Understanding and the Facts." *Philosophical Studies* 132(1): 33–42.

Elgin, C. Z. 2017. *True Enough*. Cambridge, MA: MIT Press.

Giere, R. 2006. *Scientific Perspectivism*. Chicago: University of Chicago Press.

Green, S., and Batterman, R. 2017. "Biology Meets Physics: Reductionism and Multi-Scale Modeling of Morphogenesis." *Studies in History and Philosophy of Science Part C: Studies in History and Philosophy of Biological and Biomedical Sciences* 61: 20–34.

Grimm, S. 2006. "Is Understanding a Species of Knowledge?" *The British Journal for the Philosophy of Science* 57(3): 515–535.

Grimm, S. 2008. "Explanatory Inquiry and the Need for Explanation." *The British Journal for the Philosophy of Science* 59(3): 481–497.

Hempel, C. 1965. *Aspects of Scientific Explanation*. New York: Free Press.

Kadanoff, L. P. 2000. *Statistical Physics: Statics, Dynamics, and Renormalization*. Singapore: World Scientific.

Kadanoff, L. P. 2013. "Theories of Matter: Infinities and Renormalization." In *The Oxford Handbook of Philosophy of Physics*, edited by Batterman, R. W., 141–188. Oxford: Oxford University Press.

Kaplan, D. M. 2011. "Explanation and Description in Computational Neuroscience." *Synthese* 183(3): 339–373.

Kaplan, D. M., and Craver, C. F. 2011. "The Explanatory Force of Dynamical and Mathematical Models in Neuroscience: A Mechanistic Perspective." *Philosophy of Science* 78(4): 601–627.

Khalifa, K. 2012. "Inaugurating Understanding or Repackaging Explanation?" *Philosophy of Science* 79(1): 15–37.

Khalifa, K. 2013. "The Role of Explanation in Understanding." *The British Journal for the Philosophy of Science* 64(1): 161–187.

Kim, J. 1994. "Explanatory Knowledge and Metaphysical Dependence." *Philosophical Issues* 5: 51–69.

Kvanvig, J. 2003. *The Value of Knowledge and the Pursuit of Understanding*. New York: Cambridge University Press.

Kvanvig, J. 2009. "Responses to Critics." In *Epistemic Value*, edited by Haddock, A., Millar, A., and Pritchard, D., 339–353. New York: Oxford University Press.

Lange, M. 2014. "On 'Minimal Model Explanations': A Reply to Batterman and Rice." *Philosophy of Science* 82(2): 292–305.

Lipton, P. 2009. "Understanding Without Explanation." In *Scientific Understanding: Philosophical Perspectives*, edited by de Regt, H. W., Leonelli, S., and Eigner, K., 43–63. Pittsburgh: University of Pittsburgh Press.

Longino, H. 2013. *Studying Human Behavior: How Scientists Investigate Aggression and Sexuality*. Chicago: Chicago University Press.

Massimi, M. 2018. "Perspectival Modeling." *Philosophy of Science* 85(3): 335–359.

Meier-Schellersheim, M., Fraser, I.D.C., and Klaushcen, F. 2009. "Multi-Scale Modeling in Cell Biology." *Wiley Interdisciplinary Review of Systems Biology Medicine* 1(1): 4–14.

Mitchell, S. 2009. *Unsimple Truths: Science, Complexity, and Policy*. Chicago: University of Chicago Press.

Mizrahi, M. 2012. "Idealizations and Scientific Understanding." *Philosophical Studies* 160(2): 237–252.

Morrison, M. 2011. "One Phenomenon, Many Models: Inconsistency and Complementarity." *Studies in History and Philosophy of Science Part A* 42(2): 342–351.

Morrison, M. 2015. *Reconstructing Reality: Models, Mathematics and Simulations*. Oxford: Oxford University Press.

Nozick, R. 1981. *Philosophical Explanations*. Cambridge, MA: Harvard University Press.

Potochnik, A. 2007. "Optimality Modeling and Explanatory Generality." *Philosophy of Science* 74(5): 680–691.

Potochnik, A. 2015. "Causal Patterns and Adequate Explanations." *Philosophical Studies* 172(5): 1163–1182.

Potochnik, A. 2017. *Idealization and the Aims of Science*. Chicago: University of Chicago Press.

Qu, Z., Garfinkel, A., Weiss, J. N., and Nivala, M. 2011. "Multi-Scale Modeling in Biology: How to Bridge the Gaps Between Scales?" *Progress in Biophysics and Molecular Biology* 107(1): 21–31.

Rice, C. 2013. "Moving Beyond Causes: Optimality Models and Scientific Explanation." *Noûs* 49(3): 589–615.

Rice, C. 2016. "Factive Scientific Understanding Without Accurate Representation." *Biology and Philosophy* 31(1): 81–102.

Rice, C. 2018. "Idealized Models, Holistic Distortions, and Universality." *Synthese* 195(6): 2795–2819.

Rice, C. 2019. "Models Don't Decompose That Way: A Holistic View of Idealized Models." *The British Journal for the Philosophy of Science* 70(1): 179–208.

Rohwer, Y., and Rice, C. 2013. "Hypothetical Pattern Idealization and Explanatory Models." *Philosophy of Science* 80(3): 334–355.

Rohwer, Y., and Rice, C. 2016. "How Are Models and Explanations Related?" *Erkenntnis* 81(5): 1127–1148.

Salmon, W. 1984. *Scientific Explanation and the Causal Structure of the World*. Princeton: Princeton University Press.

Strevens, M. 2008. *Depth: An Account of Scientific Explanation*. Cambridge, MA: Harvard University Press.

Strevens, M. 2013. "No Understanding Without Explanation." *Studies in History and Philosophy of Science Part A* 44(3): 510–515.

Trout, J. D. 2007. "The Psychology of Scientific Explanation." *Philosophy Compass* 2(3): 564–596.

Weisberg, M. 2007. "Three Kinds of Idealization." *The Journal of Philosophy* 104(12): 639–659.

Weisberg, M. 2013. *Simulation and Similarity: Using Models to Understand the World*. New York: Oxford University Press.

Wimsatt, W. C. 2007. *Re-engineering Philosophy for Limited Beings: Piecewise Approximations to Reality*. Harvard, MA: Harvard University Press.

Woodward, J. 2003. *Making Things Happen: A Theory of Causal Explanation*. Oxford: Oxford University Press.

Worrall, J. 1989. "Structural Realism: The Best of Both Worlds?" *Dialectica* 43(1–2): 99–124.

6 Representationalism in Measurement Theory. Structuralism or Perspectivalism?

J. E. Wolff

1 Introduction

Paradigm shifts, conceptual revolutions, or even just multiple alternative models of ostensibly the same natural phenomenon, system, or entity pose a severe challenge to traditional scientific realism. A standard scientific realist expects that our theories and models correspond to the relevant features of the natural world they are meant to represent, or that they at least aim to do so. As far as the standard realist is concerned, at most one such model will correspond to the way the world actually is; so how can more than one model enjoy predictive and other empirical successes?

In this chapter I look at two contemporary forms of scientific realism, each of which departs in crucial respects from the standard scientific realist: structural realism and perspectival realism.[1] Both take seriously the challenge of a plurality of models and theories, but they wish to retain key elements of scientific realism, such as a commitment to a correspondence between scientific representations and the world, and to the idea that science makes progress. Despite these shared commitments to realism, perspectival and structural realism offer substantially different responses to the challenges that arise from a plurality of models. After laying out the differences between the two views in section 2, I use models of measurement as a type of scientific representation to illustrate the strengths and weaknesses of structural and perspectival realism. I conclude that, at least for meta-sciences like measurement theory, structural and perspectival realism might be complementary.

2 Realism: Structural and Perspectival

Both structural realists (Worrall 1989; Ladyman 1998; French 2014) and perspectival realists (Giere 2006; Massimi 2012; Teller 2017; Teller, this volume) want to address challenges arising from the plurality of scientific representations of ostensibly the same phenomenon or subject matter while maintaining a broadly realist outlook on science.[2] The plurality of models is a challenge for the requirement of literalness of scientific representation endorsed by traditional scientific realists. The requirement of literalness is typically articulated from the syntactic view of theories and amounts to the claim that theoretical terms do not in general differ in their semantics from observational terms. The debate I focus on here, by contrast, is played out against the backdrop of the semantic view of theories, which takes *models* as paradigmatic scientific representations. Whereas a traditional realist would be inclined to hold that a successful representation of a phenomenon means that there is a close correspondence between elements of the model and elements of the phenomenon represented, both structural and perspectival realists recognize that models contain features that do not readily correspond to features in the phenomenon represented. The fact that sometimes more than one model can be used to represent the same phenomenon provides particularly strong evidence for this lack of literal correspondence between model and phenomenon. The question for both structural and perspectival realists is how to respond to the plurality of models while retaining a commitment to realism. Structural realists focus on the *commonalities* among different representations and models, whereas perspectivalists emphasize the *differences* between models.

Structural realists suggest that we should focus on what is common to competing (successful) representations and that this commonality is structural. While each model will differ from the others in some way, all models of the phenomenon will have certain structural similarities. Our task is to identify these structural similarities, which is often done by finding transformations between models that leave particular features invariant. According to structural realism, what we learn about the world from these different models is confined to the structural similarities they share. Some structural realists want to take this epistemic view further and conclude that the world itself contains nothing but structure, but for present purposes I shall be concerned only with epistemic structural realism, not ontic structural realism (for the distinction, see Ladyman 1998).

For a plurality of models of ostensibly the same entity/system, this will mean that structuralists will only take features present in *all* models as *representational*, that is, only those features that are shared between the different models will count as relevant to the question of truth-qua-correspondence, whereas features pertaining only to some models will be regarded as artifacts of the representation. Structuralists will further add that what is shared between the different models are *structural* features, which are contrasted with haecceitistic or quidditistic differences among models. Structural features typically include relations among the elements of the model which remain invariant even as we "swap" or "replace" particular elements. For example, representations of particles that differ only with respect to *which* particle in an ensemble of identically prepared particles has a given property will be regarded as only having haecceitistic difference. Such haecceitistic differences, according to the structuralists, are not up for evaluation with respect to which of them "gets it right"; instead they are merely artifacts of representation. Models that differ from the original model only in assigning different "labels" to the particles are structurally similar to the first one, and there is nothing to choose between them.[3] Structural realism reduces the plurality of representations by treating many representations as equivalent.

The relational character of structure is contrasted, on the one hand, with haecceitistic and quidditistic differences and, on the other, with nature and ontology. In both cases, the idea seems to be that the same relational structure may underpin different conceptions of the nature of the phenomenon in question, or be instantiated in models with haecceitistic or quidditistic differences. Unsurprisingly, perhaps, structuralists have focused much of their attention on the highly mathematized models used in the physical sciences. For such models, it is comparatively easy to give a characterization of the structure of the representation, and the abstraction involved in the mathematical representation makes it easier to see how the same relations can be used in otherwise different theories and models. The main challenge for structuralism is to develop a notion of structure that is both substantive enough to be controversial while also being a plausible candidate for what is in fact preserved across different theories/models.

Perspectivalists take a rather different approach to the plurality of models. Instead of focusing on the commonality among different models, perspectivalists regard each model as a complementary perspective on the same phenomenon. Unlike the structuralist, who limits what we should take as corresponding to the world in our models to what is (structurally) common to them, the perspectivalist takes differences between models as (potentially) informative about the world. Perspectivalists reject the idea that we ever approach the natural world independent of taking a particular perspective. Representation is inevitably perspectival; there is no view from nowhere (Giere 2006; van Fraassen 2008).

Moreover, not all perspectives are easily compatible. Notoriously, water is described as a viscous fluid by fluid dynamics and as a collection of particles by statistical mechanics.[4] These two descriptions seem to be in direct conflict, attributing contradictory properties to the same entity. Traditional realists would be inclined to insist that at least one of these models must be mistaken: it simply does not correspond to the nature of water. Structural realists might try to retreat to merely structural features of each model, but it is not obvious how that is going to resolve the difficulty in this case. Perspectivalists, by contrast, would like to retain both models as offering important insights into the nature of water. Neither is to be given up in favor of the other. Instead both models say something true about water, something that would be lost if we chose only one perspective as the uniquely true perspective. Whether there is nonetheless room for realism from a perspectivalist standpoint will depend on how perspectivalists can characterize the relationship between the two or more apparently inconsistent models (Giere 2009; Massimi 2018b).

Unlike structural and traditional realism, then, perspectivalism seems to be committed to a form of unavoidable pluralism. One question for perspectivalists is whether this pluralism is confined to our knowledge and representation of the world or whether it extends to the world, that is, whether the plurality of perspectives reveals that the world itself is somehow ontologically pluralistic. Especially the latter view seems to be difficult to reconcile with core commitments to realism. Realists are typically committed to a realist semantics for scientific representation, an optimistic epistemic outlook on scientific success and progress, and a picture of the world as uniquely structured in a certain way (Psillos 1999). This final point matters if we are to make sense of realism as being committed to the idea that our scientific theories correspond to what the world is like in its own right. If ontology is radically pluralistic, perhaps in the sense that entities only exist insofar as they are represented in a certain way, then this would seem to undermine a basic commitment of scientific realism. Perspectivalists in the philosophy of science do not typically wish to embrace this radical ontological departure from standard scientific realism (Massimi 2012).

The main challenge for perspectival realism, then, is to make sense of the idea that each perspective captures something true about the phenomenon in question while maintaining that these perspectives shed light on the same phenomenon or entity. This claim suggests a notion of perspectival truth that requires clarification and defense, since it seems difficult to reconcile the pluralism inherent in perspectivalism with the idea that claims about the world are either true or false. Some claims, it seems, would be true according to one perspective, yet false according to another (see Massimi 2018a for a qualified defense of perspectival truth). Moreover, something needs to be said about why it is that the different perspectives contribute something epistemically valuable to inquiry, while nonetheless remaining distinct and possibly irreconcilable. Even if the pluralism is confined to our knowledge or representation of the world, most realists would also be uncomfortable with the idea that our knowledge is always confined to perspectival knowledge only (Chakravartty 2010).

Perspectival and structural realism, then, differ in their approach to scientific representation. To assess the strength and weaknesses of each as realist approaches to scientific representation, I will now turn to measurement theory. Measurement theory addresses the question how numerical representations of empirical attributes and phenomena of interest are possible. Any form of realism about such representations will want to insist that there are some constraints on which representations qualify as adequate representations of the relevant attributes.

Measurement theory is of particular interest for the comparison of perspectival and structural realism for two reasons. First, measurement theory is not a first-order science in the manner of physics or biology; its subject matter is not a specific class of phenomena or aspect of the natural world. Measurement theory, at least as it is understood today, is a meta-science that studies the mathematical formalism used to represent measurements. What we can learn from it may hence be quite different from the conclusions we draw from case studies of models in particular sciences. Second, since measurement theory explicitly deals with a certain type of scientific representation, it seems especially appropriate to ask what structural and perspectival realists might have to say about it. In the next section I will present some problems for a literalist reading of measurement representations, which I interpret as being akin to traditional scientific realism. In sections 4 and 5 we will see how structural and perspectival realism can be combined to provide a better understanding of measurement representations.

3 Representationalism and Literalism in Measurement Theory

Measurement theory was not always a meta-science. Especially in the first half of the 20th century, the study of measurement and quantities was considered part of physical theorizing. Many important contributions to measurement theory were made by physicists, often as part of working out the foundations of physics (Tolman 1917; Campbell 1920; Bridgman 1927). The idea behind these approaches was that measurement theory was supposed to give an accurate account of physical quantities. Physical quantities were thought to be unique in permitting numerical representation, and the question was which features of these attributes made them numerically representable.

Early axiomatizations of measurement focused on the idea that quantitative attributes were numerically representable *because* they were additive (Helmholtz 1887/2010; Hölder 1901). We can both order objects of a domain by length (from shortest to longest) and concatenate objects in the domain in such a way that the combined object has the "sum" of the lengths of the two concatenated objects. Lengths, masses, and other paradigmatic physical magnitudes can be "added" in (almost) the way numbers are added. The natural conclusion for many thinkers was that quantities can be given numerical representations in virtue of being additive. Additivity was thereby made into a necessary condition for being a quantitative attribute. These early axiomatizations for quantities contained two types of axioms: axioms governing the ordering of objects and axioms governing additivity. These axioms were thought to constrain how numbers could be assigned to objects, or perhaps they were understood as something like conditions for the possibility of numerical assignment.

This approach to the question of how numerical representations of attributes are possible is characterized by a form of "literalism": it is possible to represent attributes numerically if and only if there is a direct correspondence between features of the attribute and features of the numbers. Moreover, one such feature, additivity, was selected as a necessary and sufficient condition for all numerical representation. For physical attributes, additivity had to be demonstrated empirically, by finding suitable concatenation operations for objects instantiating the attribute in question. The apparent direct correspondence between the operation of placing rods end to end, or placing weights in the same pan of a beam balance, and arithmetical addition operations on numbers was understood to be the key to the numerical representation of attributes like length and mass.

While additivity seems to fit nicely as a criterion for some paradigmatic physical quantities like mass and length, it does not fit neatly for all physical quantities. There are two types of problems. First, not all physical attributes seem to be additive in the sense that combining objects with different magnitudes of these attributes results in an increased magnitude of the attribute that could be interpreted as the sum of the two contributing magnitudes. Density and temperature are typical examples of this. Mass density is understood as mass per volume. Both mass and volume are additive quantities and hence fall squarely into the physical measurement paradigm. But while the masses and volumes of appropriately concatenated objects will increase in such a way as to form the sums of the respective masses and volumes, the same is not true for density. Fluids of different densities will typically form uniform density layers (e.g., when trying to mix honey and milk) instead of combining or produce a mixture of intermediate density somewhere between the two starting densities. Similarly, if we mix two fluids of different temperature, say coffee and milk, the resulting fluid does not have a temperature that is the "sum" of the two contributing temperatures but instead an intermediate temperature.[5]

The second type of problem is due to the operationalism built into many versions of the additivity paradigm as a result of its commitment to literalism. Additivity of an attribute is linked to the availability of a concatenation operation for objects instantiating the attribute, which means this approach rules out attributes for which no concatenation operation is available and attributes for which no *unique* concatenation operation is available.

Concatenation operations do not seem to be available for temporal intervals (except perhaps for the special case of adjacent intervals), yet we do think that time is numerically representable and indeed in some sense additive. The problem here is simply that we cannot manipulate events and intervals as easily as we can manipulate certain kinds of physical objects. Even in the case of physical objects, our ability to concatenate them is limited: we assume that the masses of planets behave in a manner comparable to that of pebbles, even though we cannot concatenate the former in the same way we concatenate the latter.

On the other hand, some quantities seem to have more than one "natural" concatenation operation. Compare, for example, electrical resistance in series and parallel circuits. Resistors connected in series yield additive resistance in a straightforward way: the total resistance in the circuit is just the sum of the resistance of each resistor. Resistors connected in parallel do not yield additive resistance, but yield the reciprocal of resistance: the total reciprocal resistance is the sum of the reciprocal resistance of each resistor. Neither parallel nor series circuits are more natural than the other, yet in both cases we seem to end up with an additive quantity: resistance and its reciprocal. The two quantities seem so closely connected that even distinguishing them seems somewhat misleading. Instead it looks like there are just two different ways of concatenating resistors, and either way of doing it yields a total resistance measure that is additive. There is no unique way of combining resistors in a circuit that yields an additive representation; instead there are two.

A similar sort of problem can be generated for the case of length. While we ordinarily assume that the natural way to concatenate lengths is to place rods end to end in a straight line, Brian Ellis (1966) showed that placing rods at right angles to each other also yields an additive representation of length, just not the one we find familiar. While Ellis's example might seem contrived, it is very difficult to say why we should prefer our standard concatenation of length to his unconventional one, other than sheer familiarity. The concern for the additivity paradigm is that the straightforward link between a natural concatenation operation and a numerical representation of the attribute featuring the addition operation breaks down.

The additivity paradigm is motivated by a form of *literalism* about numerical representations of quantities: quantitative attributes are numerically representable because, under concatenation, objects with that attribute behave like numbers with respect to addition. Numbers correspond to objects, and addition between numbers corresponds to concatenation between objects. If there is either no plausible way of concatenating the relevant objects (e.g., temporal intervals or planets) or if there is more than one plausible way of doing so (e.g., rods or resistors), then this literal interpretation becomes doubtful. There is no longer a unique, natural correspondence between the manipulation of objects (and thereby indirectly the magnitudes of quantities) and the numbers.

The literalism of the additivity paradigm is, hence, rather restrictive. While there are some physical quantities that satisfy the strict requirements of additivity (at least in a limited domain), even among physical quantities there are problem cases. For sciences other than physics, the problem is far more severe: in sciences like psychology, no attributes of interest seem to have additive structure or be amenable to concatenation. Unsurprisingly, psychologists like S. S. Stevens (1946) rejected the additivity paradigm and proposed instead that measurement simply meant the numerical representation of attributes according to some rule or other. This notion of measurement strikes many as too weak and too easily achieved (see Michell 1999 for a detailed critique of this and related notions of measurement in psychology). The question is, therefore, whether it is possible to free measurement representations from the shackles of literalism without giving up on the idea that numerical representations of attributes reflect something about the nature of the attributes thus represented.

The radical literalism of the additivity paradigm is akin to the view standard scientific realists take with respect to scientific representation in general. The standard realist expects that features of the representation correspond (literally!) to features of the phenomenon or entity represented and, conversely, they require of a representation that it captures the features of the represented entity. A close correspondence between features of the representation and features of the represented entity is what makes for successful scientific representation. This is the reasoning behind the additivity paradigm as well. Numerical representations are additive and, hence, we want to be entitled to infer that attributes represented numerically are also additive. Conversely, if a representation were to lack key features of an attribute, such as its additivity, the representation would be inadequate. For standard realists, this kind of literalism is part of what it means to be a realist.

In the following section, I will look at the representational theory of measurement (RTM), which arose in response to the problems with the additivity paradigm. I shall first show that RTM looks like a form of structural realism about representation. In section 5, we will see that this structural realism needs to be combined with perspectivalism.

4 Structural Realism in the Representational Theory of Measurement

Today measurement theory is a mathematical framework that describes the conditions under which numerical (and, more generally, mathematical) representations of attributes are possible. The most developed framework of this kind is the representational theory of measurement, which describes measurement as a representation of empirical relational structures by numerical relational structures (Krantz, Suppes, Luce, and Tversky 1971, 9). Even contemporary alternatives to representationalism, for example (Domotor and Batitsky 2008), share this highly mathematical character and do not proceed from within a particular science. A great advantage of RTM is that it describes a range of different types of structures axiomatically and shows what type of numerical representations are possible for these structures. The representationalist theory thereby incorporates a key feature of Stevens's permissive approach to measurement, namely the idea that *different* features of numerical structures can be used to establish a mapping between an empirical structure and a numerical structure. There is no need for such mappings to be confined to additive structures. Additive structures become merely a type of empirical structure that can be numerically represented.

To provide such an axiomatic framework, RTM first provides axioms for various types of structures.[6] A structure is here simply a set with relations and operations defined on it. The exact nature of the relations and operations is specified by the axioms. Crucially for RTM, both numerical structures (e.g., the real numbers, ordering, and addition) and empirical structures (e.g., a set of weights when ordered and concatenated using a beam balance) might satisfy these axioms. By characterizing structures in this abstract, axiomatic fashion, RTM lays the foundation for showing how a mapping from the empirical structure[7] to the numerical structure is possible. Such a mapping will typically be a *homomorphism*, that is, a structure-preserving map. As we represent an empirical structure using a numerical structure, the numerical structure will reflect *structural* features of the empirical structure. According to RTM, this preservation of structure is the key to understanding measurement representations. Much of RTM then proceeds to show, in a mathematically rigorous way, what kinds of representations are possible for different types of empirical structures.

To do so, first a representation theorem and then a uniqueness theorem are proved. The former demonstrates that if an empirical structure satisfies the axioms for a particular structure, for example an additive extensive structure, then there is a structure-preserving mapping from the empirical relational structure to a suitable numerical structure (suitable insofar as the numerical structure will also satisfy the axioms for additive extensive structures), such that certain conditions are satisfied. For additive extensive structures the following two conditions are satisfied: (i) the mapping is such that the ordering of objects in the empirical domain is reflected in the order of the numbers assigned to the objects: $a \prec b$ iff $f(a) < f(b)$; and (ii) the mapping is such that the concatenated object $a \circ b$[8] is mapped to the sum of the numerical values for a and b: $f(a \circ b) = f(a) + f(b)$.

The uniqueness theorem then shows how unique this mapping from the empirical structure to the numerical structure is by demonstrating how other mappings satisfying the same two conditions are related to our original mapping f. It turns out that for additive extensive structures, any mapping f' such that $f' = \alpha f$ for some real value $\alpha > 0$ will satisfy the two conditions given above. So once it has been established that one such homomorphic mapping from the empirical structure to the numerical structure is possible, many more such mappings are also possible, differing from the first one only by multiplication by a positive factor α. In measurement practice this is often taken to mean that we can change the unit of measurement, for example, from centimeters to inches, without losing any important information. The representational theory of measurement thereby shows which numerical representations are equivalent in the sense of being mere notational variants of each other.

While the preceding example illustrates the features of mappings for additive extensive structures, the same general method is applicable to other structures as well. Indeed, this is what most of the rest of *Foundations of Measurement* concerns itself with: various types of structures are axiomatically characterized and then shown to be representable by numerical structures to varying degrees of uniqueness. Whereas earlier axiomatizations of measurement had focused on capturing what was necessary for establishing the additive character of an attribute, RTM instead begins from the idea that measurement involves an axiomatic characterization of a measurement structure but does not put any constraints on the features such a structure might have. Once a measurement structure has been axiomatically characterized, we can then ask what kind of numerical representation of such a structure might be possible (the representation theorem) and how unique such a representation will turn out to be (the uniqueness theorem). Mass is numerically representable because massive objects stand in empirical relations of ordering and concatenation, that is, it satisfies the axioms for additive extensive structures. Temperature, on the other hand, is numerically representable because relations of congruence and betweenness hold between differences in temperature; temperature satisfies the axioms for absolute difference structures.[9] The features that make possible a representation of an attribute by numbers are structural features, as is clear from the fact that the mapping between them is a homomorphic mapping: a mapping that preserves structure.

The axiomatic characterizations of RTM are distinctively structural: the axioms characterize structures, that is, sets with relations and operations defined on them. This structural characterization turns out to be more abstract than the literalist construal of attributes as additive. An additive extensive structure, for example, is characterized by axioms describing a set with an ordering relation and a binary operation that satisfy certain conditions. The binary operation does not have to be addition, even though numerical addition satisfies the axioms. But other binary operations, like multiplication, work just as well. A consequence of this axiomatic approach is that even though the numerical structure used to represent a particular attribute may be additive in the sense of involving the addition operation, the attribute thus represented might lack a concatenation operation or might lack a unique concatenation operation. RTM can thereby explain some of the anomalies encountered under the additivity paradigm. Length and electrical resistance have additive extensive structures because they satisfy the abstract axioms specifying such structures. It turns out that they can do so in different ways depending on the empirical set-up chosen, but since the mapping is not thought to hold between a particular concatenation operation and numerical addition, instead holding in virtue of the satisfaction of the axioms, these cases are no longer anomalies under the new paradigm.

Moreover, since RTM describes a wide range of different structures, only some of which are characterized by axioms involving binary operations, RTM can allow for the numerical representation of attributes like temperature and other "intensive" quantities. RTM thereby avoids the constraints placed on numerical representation by the additivity paradigm.

The resolution of the anomalies and the inclusion of non-additive attributes is made possible by the move to a structural characterization of the target of measurement representations. Instead of literalism, which committed the additivity paradigm to the claim that measurable attributes must be additive like numbers, representationalism allows for a variety of ways in which attributes can have structures that satisfy specific axioms. Since the representation theorem shows that structures satisfying the axioms are representable by certain numerical structures (because it is possible to construct a structure-preserving map from the empirical to the numerical structure), the structural characterization is key to the representational theory. RTM assumes that what makes numerical representations possible is a structural similarity between numbers (and the relations and operations defined on them) and attributes, like mass or temperature (and the empirical relations and operations available for collections of objects instantiating them). Moreover, to decide when two numerical representations should count as notational variants of each other, RTM asks whether the two representations preserve the same structure. This is done through the uniqueness theorem, which compares homomorphic mappings to one another.[10] Structure that is invariant across different mappings is considered an objective feature of the attribute in question.

RTM can therefore be described as a form of structural realism about representation: structural commonalities among representations of the same attribute are indicative of objective or genuine features of the attribute, whereas features that vary in different representations (such as a change in unit) are to be regarded as conventional artifacts. Like structural realism, RTM assumes that there is a clear distinction between elements of the representation that correspond to features of the represented attribute and elements of the representation that are due to convention only; moreover, the features that correspond to features of the attribute are *structural* features only. Structural features are here once again *relations*, in contrast to haecceitistic features. Structural correspondence, as demonstrated through structure-preserving mappings, makes for successful representation for RTM.

5 Keeping Things in Perspective

RTM seems to be committed to a form of structural realism about numerical representation, and insofar as RTM is the most developed framework for measurement representations, this would suggest that structural realism provides an adequate account of numerical representation. Before drawing that conclusion, however, we must ask how widely accepted the representationalist paradigm for measurement theory is. Few would argue with the claim that the representationalist theory of measurement, especially as presented in *Foundations of Measurement*, constitutes the most thorough formal treatment of "measurement structures" and their numerical representation. Nonetheless, representationalism is not without critics, with a main line of criticism being whether representationalism truly amounts to a theory of *measurement* (Savage and Ehrlich 1992). This criticism seems even more pertinent given that RTM has not had as much of an impact on research practice in fields like psychology as might have been expected (Cliff 1992). While there are several aspects of measurement practice that seem to receive relatively little attention on the representationalist view, for the purposes of perspectivalism the most interesting criticism concerns the question of how we decide whether a given attribute satisfies the axioms for a particular measurement structure.

The application of the representationalist framework in any given scientific context requires three steps: one conceptual, the other two mathematical (Luce et al. 1990, 201). The first step (i) is to determine *whether* an attribute of interest satisfies the axioms for a given measurement structure. Once this has been established, the representationalist framework can then be used to show (ii) *that* a numerical representation of the attribute is possible and (iii) *how unique* that representation is. Steps (ii) and (iii) are important for establishing which scale type is appropriate for the attribute and which inferences can be drawn from the representation. The representationalist theory of measurement provides detailed proofs of representation and uniqueness theorems for wide range of axiomatically characterized measurement structures, which ensures that steps (ii) and (iii) are clearly justified in setting up a numerical representation. However, RTM has very little to say about the very first step.

To establish that an attribute of interest can be numerically represented, we need to know whether we have reason to believe that the attribute satisfies the axioms for some measurement structure. If such reasons can be found, then RTM simply provides steps (ii) and (iii). Whether such reasons can be found, however, will depend both on empirical observations and theoretical assumptions. Some of the theoretical assumptions are "inductive." Suppose we have found a means to concatenate objects systematically for a finite range of magnitudes for a given quantity and that these concatenations do indeed yield "sums." We might then wish to extend the assumption that magnitudes of this quantity are additive beyond the range for which we are able to carry out empirical concatenations. This is the type of assumption that leads us to conclude that mass satisfies the axioms for additive extensive structures, even though we have only concatenated a limited number of massive objects and even though some types of massive objects, like planets, eschew concatenation altogether. Another type of theoretical assumption concerns the dependence of quantities on other quantities, a situation commonly exploited in "indirect" or "non-fundamental" measurement. In these cases, the structure of one of the attributes is inferred from its nomic relationship to other attributes whose structure is presumed to be known. This situation is common for many measurements in physics, for example, for the measurement of temperature using the relationship between pressure, temperature, and volume. But how can we establish that such a nomic relationship indeed holds, without being able to measure each quantity independently?[11]

Much of the dispute in sciences like psychology and in other fields concerns precisely the question whether we are justified in assuming that a particular attribute indeed satisfies the axioms for a particular measurement structure. Reasons to support such a claim are never free from theoretical assumptions of the sort mentioned and can hence be contested. This theory dependence opens the door for a more perspectivalist reading of measurement representations.

The perspectivalist reading begins from the observation that an axiomatic structure by itself does not represent anything in particular. To be a representation of a particular attribute or empirical structure, the axiomatic structure needs to be interpreted. This interpretation connects aspects of the phenomenon of interest to the axioms characterizing that abstract structure. For measurement structures the interpretation will involve characterizing the phenomenon or attribute of interest as having a certain structure (van Fraassen 2008). Interpretations like these, as we have just seen, can be contested because they make theoretical assumptions.

Philosophers sometimes seem to think of interpretation as the task of finding a suitable empirical interpretation for an otherwise unspecified axiomatic structure. But while this highlights the way in which axioms leave their interpretations unspecified, this is not the way most scientists encounter the problem. Quite the reverse. Scientists typically start from an attribute or phenomenon they wish to represent numerically. The question is, *which* structural representation is appropriate? Ostensibly the same attribute (e.g., utility or temperature) or phenomenon (e.g., light or water) is given different structural characterizations in the context of different theories. For example, Bradford Skow has argued that thermodynamics only provides very weak reasons for thinking that temperature has a metric structure (either an absolute difference or a ratio scale structure), whereas statistical mechanics provides strong reasons for thinking that temperature has metric structure (Skow 2011). Which (measurement) structure we are justified in ascribing to temperature, then, depends on our theory of temperature.

This type of theory dependence looks like the theory dependence of other types of scientific representation. Recall the models of water as a fluid and as a collection of particles we briefly discussed at the beginning. Traditional realists will suggest that at most one of these models correctly represents water, whereas perspectivalists suggest that each offers an informative perspective on water. Similarly, for the case of temperature, a traditional realist will be inclined to suggest that statistical mechanics provides the correct account of what temperature is and, hence, the correct assignment of structure to temperature. By contrast, a perspectivalist will insist that we take seriously both the thermodynamic and statistical mechanical perspectives on temperature. Either way, the structural realism embedded in RTM does not tell us whether to go with traditional realism or perspectivalism on this point, since RTM does not deal with the question of how to decide which structure to ascribe to particular attributes.

Perspectival realists differ in their responses to the problem of inconsistent models (compare, for example, the difference between Giere 2006 and Massimi 2018b). In the case of different measurement structures ascribed to temperature, it is tempting to conclude that, since statistical mechanics is a more fundamental theory than thermodynamics, we should simply go with the structure ascribed to temperature by statistical mechanics. This reading seems even more compelling when we remember that in this case the two structures ascribed to temperature are not strictly speaking incompatible. After all, the metric structure ascribed to temperature by statistical mechanics is simply stronger than the mere ordinal structure implied by thermodynamics: an attribute that possesses metric structure also possesses ordinal structure. Semantically, then, we should accept the structural realists' claim that the structure of the numerical representation corresponds to the structure of the attribute. Which representation is adequate is then a question of what structure the attribute actually has. Structural realism here looks very much like traditional realism in its commitment to a correspondence between features of the representation and features of the represented attribute. The only difference between structural realism and traditional realism is that structural realism restricts this correspondence to structure only.

Perspectival realism has a different contribution to make, however. As Massimi (2012) has argued, perspectivalism contributes to the realist quest by supplying the relevant notion of justification. To be realists, not only do we need to have a realist semantics of the relevant representations, but we also need a realist epistemology that distinguishes justified from unjustified beliefs about the phenomena and entities in question.

The justification (as opposed to the aptness) of a belief, according to Massimi, is a matter of coherence with a given scientific perspective.

This notion of perspectival justification is relevant for the case of the attribution of measurement structures. The literalist additivity paradigm proceeded from the assumption that the availability of an empirical concatenation operation was both necessary and sufficient for an attribute to qualify as a quantity. No other justification could be given for representing an attribute numerically. The additivity paradigm thereby provided a universal criterion for quantitativeness, with no room for different theoretical approaches. On the representational theory, this requirement has been given up, but at the cost of leaving open how we should justify ascribing a specific structure to an attribute. Perspectival realism (of the epistemic variety) fills in the gap. Measurement structure is ascribed to an attribute from within a scientific perspective, such as thermodynamics or statistical mechanics. The ascription of a particular measurement structure to a given attribute has to cohere with the relevant theoretical background commitments and beliefs. While many scientists speculated that temperature might have a metric structure even before the advent of statistical mechanics (see Skow 2011 for discussion), it is only from the perspective of statistical mechanics that such an ascription is justified.

Since RTM is silent on the question of how to justify the attribution of measurement structures to attributes, the structural realism at work in RTM is insufficient to satisfy the realist quest for measurement representations. Structural realism only provides the realist semantics for measurement representations, since it specifies which features of the numerical representation correspond to features of the attribute. Perspectival realism is needed to complement this picture, since perspectival realism provides a notion of epistemic justification for attributions of measurement structures that makes sense of the different attributions of measurement structures by different theories.

6 Conclusion

In this chapter, I looked at structural realism and perspectival realism initially as two competing responses to the plurality of scientific representations. My focus has been on representations of measurement. I argued that literalism about measurement representations, which corresponds to a traditional form of realism about representation, is inadequate. The representational theory of measurement, which provides a thoroughgoing account of numerical representations, implies structural realism about measurement representations. While RTM avoids some of the difficulties with the literalist reading, it needs to be supplemented with epistemic perspectival realism to account for the theory dependence of our justifications for ascribing particular measurement structures to attributes of interest.

There is a broader, more speculative lesson we might learn from this case. At least in some cases, structural realism and perspectival realism are not competing realist accounts of representation but instead offer complementary pieces in the realist quest. Structural realists do well in supplying a more appropriate notion of correspondence between representation and represented phenomenon by freeing representation from literalism. Perspectival realists, on the other hand, provide an account of the justification of using different representations of ostensibly the same phenomenon or attribute. This is important since, even after all equivalent (numerical) representations have been explained by structural realism, different scientific theories still ascribe different structures to the same attribute.

Notes

1. There are non-realist versions of both structuralism and perspectivalism, which I do not have room to discuss explicitly in this chapter. These views share the respective structural and perspectival outlook on scientific representation (discussed in the next section) but without any commitment to a form of scientific realism.
2. As with any view in philosophy, there are in-house disputes among structural realists and perspectival realists. Since my main purpose here is to contrast the two different approaches, I shall set aside the finer points of disagreement in each camp.
3. The question of whether haecceitistic differences matter in the context of quantum particles is discussed in, for example, French (1989) and Huggett (1999).
4. For discussions of this example, see Morrison (1999) and Teller (2001).
5. Quantities that behave like density and temperature have sometimes been called "intensive quantities," in contrast to extensive quantities like mass and volume. The distinction between intensive and extensive quantities is not always clearly defined, nor is it uncontested. Tolman (1917, 239) defined intensive quantities to be non-additive; Suppes (1951), by contrast, replaces the additivity demand with the idea that extensive quantities are quantities that can be represented on ratio scales, whereas intensive quantities are only representable on weaker scales.
6. For example, RTM describes a structure as a (closed) additive extensive structure, by providing a set of characteristic axioms (Krantz et al. 1971, 73):

    Let A *be a nonempty set,* $\succsim$ *a binary relation on* A, *and* ∘ *a closed binary operation on* A. *The triple* ⟨A, $\succsim$, ∘⟩ *is a* closed extensive structure *iff the following four axioms are satisfied for all* a, b, c, d, ∈ A:

    1. *Weak order:* ⟨A, $\succsim$⟩ *is a weak order, i.e.,* $\succsim$ *is a transitive and connected relation.*
    2. *Weak associativity:* a ∘ (b ∘ c) ~ (a ∘ b) ∘ c.
    3. *Monotonicity:* a $\succsim$ b iff a ∘ c $\succsim$ b ∘ c iff c ∘ a $\succsim$ c ∘ b
    4. *Archimedean: If* a ≻ b, *then for any* c, d ∈ A, *there exists a positive integer* n *such that* na ∘ c $\succsim$ nb ∘ d, *where* na *is defined inductively as:* 1a = a, (n + 1)a = na ∘ a.

7. Traditionally these structures are understood as domains of concrete objects and "empirical," that is, observable qualitative relations among them. This interpretation reflects the empiricist and operationalist commitments of the

founders of RTM, but it is not the only available interpretation. Instead the structures can be understood, for example, as sets of space-time points with relations among them (Field 1980).

8. As before, the concatenated object might be understood as two physical objects combined operationally (e.g., by placing two rods end to end). Crucially, it is a mapping between an empirical or physical domain and a numerical (or mathematical) domain.

9. I will return to the question of how we are justified in ascribing a particular structure to an attribute in section 5.

10. The technical details of this comparison are a bit too elaborate to be included in the discussion here. For relevant literature, see Luce, Krantz, Suppes, and Tversky (1990), especially chap. 20.

11. For a detailed discussion of this problem in the case of temperature, see Chang (2004).

References

Bridgman, P. W. 1927. *The Logic of Modern Physics.* New York: Macmillan.

Campbell, N. R. 1920. *Physics: The Elements.* Cambridge: Cambridge University Press.

Chakravartty, A. 2010. "Perspectivism, Inconsistent Models, and Contrastive Explanation." *Studies in History and Philosophy of Science Part A* 41(4): 405–412.

Chang, H. 2004. *Inventing Temperature: Measurement and Scientific Progress.* Oxford: Oxford University Press.

Cliff, N. 1992. "Abstract Measurement Theory and the Revolution That Never Happened." *Psychological Science* 3(3): 186–190.

Domotor, Z., and Batitsky, V. 2008. "The Analytic Versus Representational Theory of Measurement: A Philosophy of Science Perspective." *Measurement Science Review* 8(6): 129–146.

Ellis, B. D. 1966. *Basic Concepts of Measurement.* Cambridge: Cambridge University Press.

Field, H. 1980. *Science Without Numbers: A Defense of Nominalism.* Princeton: Princeton University Press.

French, S. 1989. "Why the Principle of the Identity of Indiscernibles Is Not Contingently True Either." *Synthese* 78(2): 141–166.

French, S. 2014. *The Structure of the World: Metaphysics and Representation.* Oxford: Oxford University Press.

Giere, R. N. 2006. *Scientific Perspectivism.* Chicago: The University of Chicago Press.

Giere, R. N. 2009. "Scientific Perspectivism: Behind the Stage Door." *Studies in History and Philosophy of Science* 40(2): 221–223.

Helmholtz, H. v. 1887/2010. "Zählen und Messen, erkenntnistheoretisch betrachtet." In *Heidelberger Texte zur Mathematikgeschichte*, edited by Dörflinger, G. Heidelberg: Universitätsbibliothek Heidelberg. www.ub.uni-heidelberg.de/archiv/13161.

Hölder, O. 1901. "Die Axiome der Quantität und die Lehre vom Mass." *Berichte über die Verhandlungen der Königlich Sächsischen Gesellschaft der Wissenschaften zu Leipzig* 53: 1–46.

Huggett, N. 1999. "Atomic Metaphysics." *The Journal of Philosophy* 96(1): 5–24.

Krantz, D. H., Suppes, P., Luce, R. D., and Tversky, A. 1971. *Foundations of Measurement.* New York: Academic Press.

Ladyman, J. 1998. "What Is Structural Realism?" *Studies in History and Philosophy of Science Part A* 29(3): 409–424.

Luce, R. D., Krantz, D. H., Suppes, P., and Tversky, A. 1990. *Foundations of Measurement: Representation, Axiomatization, and Invariance. Volume 3.* San Diego: Academic Press.

Massimi, M. 2012. "Scientific Perspectivism and Its Foes." *Philosophica* 84: 25–52.

Massimi, M. 2018a. "Four Kinds of Perspectival Truth." *Philosophy and Phenomenological Research* 96(2): 342–359.

Massimi, M. 2018b. "Perspectival Modeling." *Philosophy of Science* 85(3): 335–359.

Michell, J. 1999. *Measurement in Psychology: Critical History of a Methodological Concept.* Cambridge: Cambridge University Press, 2004.

Morrison, M. 1999. "Models as Autonomous Agents." In *Models as Mediators*, edited by Morgan, M., and Morrison, M., 38–65. Cambridge: Cambridge University Press.

Psillos, S. 1999. *Scientific Realism: How Science Tracks Truth.* London: Routledge.

Savage, C. W., and Ehrlich, P., eds. 1992. *Philosophical and Foundational Issues in Measurement Theory.* Hillsdale: Lawrence Erlbaum Associates.

Skow, B. 2011. "Does Temperature Have a Metric Structure?" *Philosophy of Science* 78(3): 472–489.

Stevens, S. S. 1946. "On the Theory of Scales of Measurement." *Science* 103(2684): 677–680.

Suppes, P. 1951. "A Set of Independent Axioms for Extensive Quantities." *Portugaliae Mathematica* 10(4): 163–172.

Teller, P. 2001. "Twilight of the Perfect Model Model." *Erkenntnis* 55(3): 393–415.

Teller, P. 2017. "Modeling Truth." *Philosophia* 45(1): 143–161.

Tolman, R. C. 1917. "The Measurable Quantities of Physics." *Physical Review* 9(3): 237–253.

van Fraassen, B. C. 2008. *Scientific Representation: Paradoxes of Perspective.* Oxford: Oxford University Press.

Worrall, J. 1989. "Structural Realism: The Best of Both Worlds?" *Dialectica* 43(1–2): 99–124.

7 Safe-and-Substantive Perspectivism

David Danks

1 Bases for Perspectival Models

Recent years have witnessed a renaissance in "other-than-realist" approaches in the philosophy of science, whether pragmatist, instrumentalist, or one of many other -ists. Many of these approaches can be broadly understood as advocating some kind of perspectivism: roughly, the idea that scientific theories, models, knowledge, and claims are from a perspective, rather than necessarily expressing objective, universal truths. The idea that scientific claims are perspectival does not automatically imply any kind of strong relativism about science, but it does imply that there are problems with naïve realism. Of course, this high-level characterization subsumes an enormous diversity of accounts, depending on the nature of a "perspective," the reasons why science employs perspectives, the implications for scientific practice, and more. For example, a pragmatist philosophy of science might emphasize the necessary role of practical goals in our scientific theories, while an instrumentalist philosophy of science might focus on the role of measurement methods and instruments. Nonetheless, both agree that science necessarily instantiates various perspectives, rather than consisting of universal truths. Overall, much of this other-than-realist philosophy of science has emphasized either the nature of these perspectives in science or the methodological implications of such perspectives (though not always using the language of "perspectives").

In this chapter, I aim to provide a detailed account of two reasons for perspectives in science. In general, perspectivist approaches in the philosophy of science face a significant challenge, as these scientific perspectives must arguably be grounded in sources that are neither specific to the individual nor high-level banalities. If the relevant perspectives are individual specific—that is, the relevant aspects of some scientific perspective are based on particular features of particular scientists or particular research groups—then we have an "unsafe" perspectivism. Such a view implies that science itself is dependent on local, contingent properties of specific people, and so we have as many sciences as we have scientists. Hyperlocal perspectivism means that science does not provide us with a shared view of the world but rather personalized, individualized accounts. One of the hallmarks of science, however, is supposed to be exactly its ability to produce objective accounts of the world (or, at least, more objective than ordinary cognition). At the same time, if we instead ground the perspectival nature of science in high-level claims about humans and our practices, then we risk having an "insubstantial" perspectivism. While it is true that science is done by bounded humans, this observation provides no insight or guidance into the nature of scientific perspectives. If we want our philosophical frameworks to be helpful in some way, then we should insist that any such account, including perspectivist ones, should consist of more than obvious truisms.

In contrast with both of these extremes, I argue here that the influences of scientific concepts and scientific goals imply that science is necessarily, but also unproblematically, perspectival. My arguments will focus on the causes of perspectivism, and thus will not provide a definition of exactly what constitutes a perspective. Instead, this type of analysis aims to reveal some (though obviously not all) key perspectival features of science, which jointly imply that standard realist views cannot be correct. The resulting perspectivism provides specificity about the nature and impact of these factors and thereby provides substantive constraints and methodological implications for scientific practice. At the same time, this approach blurs the lines between scientific and everyday perspectives, thereby implying that perspectivism in the philosophy of science is no more problematic than perspectivism about everyday perception. More precisely, these sources of perspectivism are not unique to scientific theories, knowledge, and beliefs but rather apply to their everyday counterparts. That is, there is nothing special (with respect to these arguments) about science, and so the resulting perspectivism about science does not threaten a collapse into complete relativism (or at least, poses no more threat than we face about *all* of our beliefs and knowledge).

I begin by examining these two sources of (scientific) perspectives in more detail: concepts in section 2 and goals in section 3. For both types of influences, I focus on the ways in which particular scientists' concepts and goals impact their scientific theories, models, and knowledge. That is, my approach here employs (mostly) methodological individualism, as I largely focus on the influences of concepts and goals of particular scientists rather than the concepts and goals of scientific communities. In particular, notions such as paradigms, research programs, or similar group-level frameworks enter into this analysis through the cognition and activities of particular scientists rather than through some independent social existence.[1] Section 4 then takes up a more general discussion of the resulting scientific perspectivism, both characterizing it and showing that it blurs smoothly into more everyday, prosaic perspectivism. Thus, there is nothing to fear from (this kind of) scientific perspectivism: science does not provide an objective, universal mirror of the world, but its distortions are no more problematic than those of our ordinary, everyday perception and cognition about the world. Yes, science is perspectival, but in a safe-and-substantive way.

2 Sources of Perspectives: Concepts

Our scientific and everyday cognition is thoroughly conceptualized: our understanding of the world is almost entirely in terms of concepts[2] rather than some kind of unconceptualized, direct access to the world. There are, of course, numerous debates about whether *some* aspect of (early) perception is perhaps nonconceptual (e.g., Dretske 1981; Evans 1982; Crane 1992; Peacocke 1992, and many other papers in subsequent years), but there is no debate about whether our thinking inevitably involves concepts and conceptualized content at some point. We acquire concepts from a very young age, and those concepts and conceptual frameworks are essentially ubiquitous in our cognition. These observations are no less true for scientific cognition, though scientific concepts are often more clearly articulated and more widely shared (in some sense) within the scientific community.

A common belief about concepts, at least those of the everyday sort, is that they simply provide a compact encoding of information about the world. On this view, concepts enable us to efficiently and quickly encode relevant information about the state and structure of our environment. For example, much of the psychological literature on concept acquisition emphasizes the tight connection between environmental statistics and learned concepts for those environments. This line of research emphasizes the ways in which concepts encode environmental regularities and thereby help to identify what is relevant, anomalous, and so forth. The underlying intuition is that one major function of concepts is to convert a messy, complex external world into cleaner, relatively more tractable cognitive representations.

This way of thinking about concepts suggests that they largely play a filtering role. In general, more compact representations will almost always involve a loss of information relative to the original, but the talk of statistical encoding (among other features) suggests that the loss might involve only irrelevant information. If that were correct, then concepts could be understood as a non-distorting information filter that provide a "mirror" of the world (at least, for all of the information that made it through the encoding process). Unfortunately, this line of thinking is mistaken: concepts do not simply filter information about the world but rather actively influence and transform that information. That is, concepts distort the world (when compared to a mirror) and so constitute a substantive element of a perspective. Although there are many phenomena to which one could point, I focus here on only three.

First, consider the phenomenon of categorical perception (Harnad 1987; Goldstone and Hendrickson 2010): at a high level, instances that are close to a category boundary are perceived (in conscious cognition) as further from that boundary than they actually are. More precisely, when some instance X is understood as falling under the concept C, then the perception of X is shifted toward the centroid (in the relevant feature space) of C. Perhaps the best-known instance of categorical perception arises in phonemic discrimination. Many pairs of phonemes in a language will differ on only one acoustic dimension; for example, the phonemes /r/ and /l/ form such a pair. Individuals who learn at a sufficiently young age (Eimas, Siqueland, Jusczyk, and Vigorito 1971) a language in which these are distinct phonemes lose the ability to "hear" sounds that are intermediate between these phonemes. Instead, they hear intermediate sounds as something much closer to the central phoneme sound. Moreover, these unconscious discriminations are resistant (though not immune) to alteration through training (Strange and Dittmann 1984). More importantly for our present purposes, "hearing" a sound as a particular phoneme involves a distortion of the world: the experienced sound is simply not a mirror of the acoustic properties of (that part of) the world. Rather, categorical perception involves changes to the closeness relations of various stimuli and, more generally, a shift in the perceived "location" in perceptual space (Goldstone 1994; Livingston, Andrews, and Harnad 1998). This type of categorical perception is not limited to phonemic discrimination but rather arises for a very wide range of concepts, arguably every concept that has a perceptual component. Our concepts and categories have been shown to (directly) influence our experience of the world in perceptual modalities such as visual perception (Livingston et al. 1998), as well as more complex, not purely perceptual concepts (Etcoff and Magee 1992).

This focus on "distortions" due to primarily perceptual concepts might seem irrelevant to most of our scientific cognition. As the second example shows, however, expertise can play a significant role in the concepts we form and therefore the ways that the world appears to us. As just one example, Medin, Lynch, Coley, and Atran (1997) showed that the plant-related concepts of park maintenance workers are quite different from the plant-related concepts of taxonomists. That is, people whose job required a focus on the ecological niche of park trees had significantly different concepts than people whose job required a focus on genetic or biological relationships. Moreover, those different concepts made a behavioral difference in reasoning, inference, and descriptions, and are not intertranslatable in any straightforward way; they carve up the world in different terms. Of course, while concepts do more than just represent summary statistics, they also do have that representational function. Thus, as someone gains more experience and expertise in a domain, her concepts can significantly shift as she learns more about the relevant summary statistics. This conceptual change would not necessarily be an issue, except that those same concepts influence both basic perceptions (see the previous paragraph) and more complex, conscious cognitions (Cohen, Dennett, and Kanwisher 2016). Thus, conceptual change can have quite wide-ranging and hard-to-predict impacts on other concepts. Of course, all of these observations do not mean that concepts are somehow non-veridical (Cohen 2015), but rather that the content of the concepts—both perceptual and cognitive content—is not what we might expect. In particular, that content does not provide a simple mirror of the world in the way assumed by simple realist models, whether of perception, cognition, or even science.

As a third demonstration of the perspectival nature of concepts, consider our episodic memories of particular experiences, such as my memory of eating breakfast this morning. A common view of episodic memories (at least, within people who do not study memory) is that they involve relatively direct recall of the earlier events. Of course, that recall is subject to many types of noise and error, and so our memories need not be particularly accurate. However, this noise is (on the common view) largely independent of the content of the memories, though it can depend on the circumstances in which the memory is initially encoded (e.g., emotionally laden events are often thought to be more likely to be correctly encoded and recalled). However, there is now substantial research demonstrating that episodic memories involve a process of reconstruction, not simply one of recall. For example, our memory of an event at some past time is "recalled" using the concepts that we have now (Schacter, Norman, and Koutstaal 1998; Conway and Pleydell-Pearce 2000). Hence, if our concepts change between the time of the events and the recall time, then the memory will shift along with the conceptual change. Alternately, if we are asked to recall whether we have previously seen particular images, then we will make more errors on previously unseen images that are close to actual prior images when both are not too far from the concept centroid (Koutstaal and Schacter 1997). The distortions in our episodic memories are not ubiquitous or uniform but rather depend on relatively fine-grained details of our concepts at that later moment in time.

One might object that these roles of concepts fall under ordinary, garden-variety theory-ladenness of observation. Philosophers of science long ago became used to the idea that our theories, including our concepts, influence our observations (Hanson 1958; Kuhn 1962). For example, we look at a needle deflection and instead "see" an atom undergoing radioactive decay, or we look through a microscope at some squiggles and "see" a cancer cell. The observations that we qua scientists record and use in our scientific practices are themselves conceptualized by the scientific concepts in our theories. Hence, this section might appear to be much ado about nothing. In response, we should first note that at least one conception of theory-ladenness of observation does not fit with the phenomena described in this section: namely, scientific concepts (and the theories from which they are built) cannot simply act as a "filter" that identifies certain features or properties as bundled together in a concept and thereby ignores the others. In the examples above, the act of "seeing" a cancer cell is not simply a categorization judgment; rather, as with other kinds of concepts, we should expect that this act will also distort (relative to more objective measures) the perceptions of the squiggles in a top-down way.

Observations are not merely theory-laden but rather are theory-*shaped* or theory-*distorted*. Our understanding of human concepts implies that our scientific observations should be pulled toward the centroid of the relevant concepts; shaped by the functions for which we use those concepts; and potentially, even unknowingly, revised over time as the scientific concepts shift. More generally, the role of concepts that I have outlined in this section is significantly more active than one often finds in discussions of the theory-ladenness of observations. At the same time, I grant that everything I have written to this point is consistent with a philosophical account of theory-ladenness that is based on the fact that we humans perceive the world in ways that are distorted (depending on our concepts), and so scientific perception is distorted. However, such philosophical accounts are often used to argue for a broader type of relativism or incommensurability (Kuhn 1962; Feyerabend 1975; Longino 1990) and so contrast with the larger, non-relativist view that I develop here (see section 4).

3 Sources of Perspectives: Goals

A second set of causes of the perspectival nature of science—again not constitutive of those perspectives—is the goals and intended functions or tasks of scientists. That is, I contend in this section that our cognition about the world is deeply shaped, and arguably distorted in key ways, by the goals that we have or the tasks that we believe we will need to perform in the future. One might immediately object that this proposal cannot be right, as goals should only enter into our cognition (whether scientific or not) when we are engaged in reasoning and decision-making. This "standard view" holds that our learning and conceptualization of the world aim solely to reflect the structure, both causal and statistical, of the learning environments. Of course, as we saw in the previous section, concept learning can lead to perceptions that are distorted in various ways, but the standard view holds that those distortions are not driven by goals. That is, the core content of our concepts should, on this view, be goal-free. Many standard cognitive models of learning embody this standard view: Bayesian learning algorithms, neural networks, and most computational models of learning all mirror environmental statistics without regard to goals.[3] On this view, goals enter into cognition only after we have learned concepts that roughly mirror the world.

While this standard view is appealing in many ways, it is arguably not normatively justified. If a cognitive system, whether human or other, has to interact with its world, then the ultimate measure of its learning will be whether the learned content enables the system to succeed. For example, if the system should select option *A* anytime the perceived object is between 0 cm and 2 cm long, then there is no extra value to encoding the precise length, rather than only the fact that the object falls into the relevant interval (Danks 2014; Wellen and Danks 2016). Moreover, if the system exhibits any noise in its decision-making processes, then there can actually be an incentive to "misperceive" the object as being further from the decision boundary than it actually occurs, as that misperception will increase the likelihood of the system answering correctly (Hoffman, Singh, and Prakash 2015; O'Connor 2014). For example, an object that is 1.9 cm ought to be perceived as closer to 1.5 cm if the decision boundary is set at 2 cm, though that same shift ought not occur (to a noticeable degree) if the decision boundary is set at 10 cm. More generally, there is a normative argument that cognitive systems ought to sometimes be indifferent to believing falsehoods and sometimes ought positively to believe falsehoods. For example, if some false belief fits more cleanly with our other knowledge (perhaps because of a shared structure or analogy) and that falsehood does not impair our ability to succeed at various goals, then we ought to go ahead and believe the falsehood. Of course, falsehoods or inaccuracies that impair our ability to achieve our goals (whatever those might be) ought to be rejected during learning. Nonetheless, the door is open for goals possibly having a significant impact on our *learning*, not solely our reasoning and decision-making.

In fact, the descriptive data reveal that people often do have these kinds of inaccurate or false beliefs, exactly when they do not impact our ability to achieve our goals. For example, if people are shown multiple sequences of numbers and asked to estimate which sequence has the largest (or, alternately, smallest) average value, then they learn relatively little about the sequences that are clearly goal-irrelevant (e.g., low-magnitude sequences when the goal is to learn which has the largest mean), to the point of failing to distinguish between sequences that are easily distinguishable when they are goal-relevant (Wellen and Danks 2014; Wellen 2015). In these studies the only variation between people is what goal they were provided in the experimental cover story, and so that is the only available explanatory factor for the significant differences in *learning*, not simply reasoning. Alternately, if people have the goal of "learn to control a dynamical system," then they learn relatively little about the underlying causal structure governing the system, even though they have no trouble with that given the goal of "learn the causal structure" (Hagmayer, Meder, Osman, Mangold, and Lagnado 2010). Many more examples of this type can be found in the empirical literature (Ross 1997, 1999, 2000; Markman and Ross 2003).

Moreover, there are also cases of goal-determined learning of falsehoods, not just failures to learn. For example, Feltovich, Spiro, and Coulson

(1989) showed that many medical doctors (at that time) had incorrect beliefs about the causal direction between *heart size* and *heart strength* in congestive heart failure: the doctors believed that the causal connection was *size → strength*, but the actual physiology is *size ← strength*. Moreover, the true causal direction was known at the time of Feltovich, Spiro, and Coulson's study; the relevant information was readily available to the medical doctors. However, the false belief had no practical impact given the medical technologies and interventions available to doctors at the time. And there were positive reasons for doctors to believe the falsehood, as it fit cleanly with their knowledge about other muscles in the human body. Hence, if doctors have the goals of diagnosis and treatment while minimizing or reducing cognitive effort (given the complexity of the domain), then they ought and do learn a falsehood. Alternately, if people are charged with manipulating the world to bring about an outcome, then they will often systematically mislearn the causal structure of the world, though in exactly the right way to minimize the probability of incorrect action (Nichols and Danks 2007). Again, we have a case in which the goals influence the learning in deep ways.

If all of this is correct, then we should expect our *scientific* goals to impact our *scientific* learning, whether to yield various inaccuracies (which are goal-irrelevant) or perhaps even justifiable falsehoods. One response would be to argue that scientists share a single goal—namely, to discover the truth—and so these observations about everyday learning are unproblematic: there will be no variation in what is learned (since we all have the same goal), and we ought not learn falsehoods (since that would fail to satisfy the goal). However, this single goal cannot actually be the guide to scientific learning, as we have no way of directly assessing whether we are moving closer or further from it; we have no Archimedean point from which to assess the truth or truth-aptness (or whatever concept one prefers) of our scientific theories (Kitcher 1993; Danks 2015). Of course, the scientific community could perhaps have a single goal guiding all of their inquiries, though that goal cannot be "discover the truth." Once we rule out this overarching truth-centric goal, though, then it is hard to imagine what that single goal might be.

Science instead arguably proceeds through convergence, as we employ multiple methodologies in the hope that they will imply the same theory, the same concepts, or the same representations of the world. When our multiple methods seemingly lead to the same answer, then we conclude that we must be tracking *something* truthful about the structure of the world. Hence, our ability to "objectify" our measurements and conclusions might be taken as evidence that goals are not actually playing a significant role in our scientific learning. As noted earlier, of course, the empirical phenomena discussed in this section do not imply that we should *always* be learning falsehoods; sometimes, the best thing to learn might be the truth (at least, in experimental settings where we can talk sensibly about knowing the truth). The challenge is that we do not know a priori whether we are in such a circumstance. Perhaps our goals either should or do instead lead us toward biased or distorted learning. Our mixture of scientific goals—prediction, explanation, discovery of unobservables, and so on—might be best satisfied by learning the truth (whatever that exactly means for the world), but we have no particular reason to expect that at the outset, nor do we have any way to test it. Moreover, the existence of a single best (scientific) theory is not informative in this regard: for any given goal or mixture of goals, there will typically be a unique theory that optimizes performance relative to that goal or goals (Danks 2015). We know in advance that there will be a best theory relative to our scientific goal(s); we just do not know whether it will be the correct (or true) one.

These considerations seem to point, though, toward a reductio against my conclusion: (1) scientists clearly exhibit a diversity of goals in terms of what they are trying to explain or predict, even within a scientific domain; thus, if (2) different goals imply different concepts and theories, then we should expect diversity of scientific concepts; but (3) we are able to communicate and debate with one another in scientific contexts, and so we must not have this kind of conceptual diversity (and hence, proposition (2) must be incorrect). However, when we look at scientific practice, we do sometimes see exactly the kinds of diversity that proposition (3) denies. For example, consider the goal of explaining how people perform certain kinds of key cognitive operations, whether concept learning, decision-making, various predictions, or other cognition. This goal is actually ambiguous between explanations that are grounded in rational justifications about the limited nature of human cognition—so-called rational process theories (Denison, Bonawitz, Gopnik, and Griffiths 2013; Vul, Goodman, Griffiths, and Tenenbaum 2014)—and those based on descriptive, empirical observations and constraints—the process models traditionally developed by psychologists to model the actual mechanisms of the mind. Crucially, scientists pursuing these two different subgoals have demonstrated exactly the predicted difficulties in communication, such as debates that seem to involve all parties talking past one another. Moreover, the core problem in the discussions between researchers with these two different goals is precisely that they do not agree about the standards for evaluating the proposals. Both sides are trying to answer questions about "how the mind actually does what it does," but one side (rational process theorists) requires normative justification for the theory and the other (traditional process or mechanism modelers) requires precise empirical validation of the model. The different goals translate directly into different learnings and therefore into different understandings of the human mind.

4 Everyday Perspectivism

Given these observations, I propose that a perspective should include (though not necessarily be constituted by) the particular concepts, goals, and thus accompanying distortions. Importantly, this characterization implies that every individual has a perspective, but perspectives are not relative to specific individuals. Two different people could have the same perspective, as long as they have the same (up to relevant noise, error, or change) concepts and goals. For example, we might plausibly think that members of a research group would likely share concepts and goals, as they work closely and presumably discuss what is meant by their terminology, and what standards or goals are relevant for their research. Since perspectives are individual-independent objects, they can be shared across many people; in fact, some measure of shared perspective is almost certainly required for certain types of debates. Moreover, perspectives can be judged against various standards, whether the goals that they contain or some other goal. If one thinks, for instance, that empirical prediction is a goal that should be part of every legitimate scientific perspective, then we can assess various putative perspectives according to that standard, even if the perspective is developed with emphases on other goals (e.g., explanatory power). In addition, this conception of a perspective implies that an individual's history, relevant sociocultural factors, measurement methods, and so forth should all be rendered irrelevant once we know their concepts and goals.[4] Of course, an individual's history matters, but on this account only inasmuch as that history leads to the individual having a particular set of concepts or because of the goals that the individual had at some earlier point in time. In particular, multiple individuals might share relevant aspects of their histories and so share some concepts and goals.

The concept- and goal-based perspectivism that I have outlined here is thus "safe" in the sense that it does not automatically lead to a descent into hopeless relativism. For most interesting scientific domains and research challenges, the practices of scientific training (which arguably homogenize the community along the lines of concepts and goals) and also people's shared cognitive architectures (by virtue of being human beings) should lead to most scientists having, in practice, relatively similar perspectives. There is little reason to think that multiple scientists' concepts or goals are so different as to imply that there are substantively distinct perspectives. Moreover, the world "gets a say" in the perspectives, as there will typically be a normatively unique (or close to unique) set of concepts and theories for a set of goals in a scientific domain, though we might not, in practice, be able to determine that set. Relatedly, our concepts are not arbitrary or ungrounded in experience but rather are learned from experience. We cannot simply invent and use whatever perspective we might want. Rather, we are significantly constrained by the world in terms of the acceptable perspectives, at least once we have specified the relevant goals (and sometimes some auxiliary concepts).

At the same time, this type of perspectivism is substantive, as it is not simply the banality that "humans do science" (and so science is done from the "human" perspective). Rather, this perspectivism is grounded in features of human cognitive processing and representations: the details of our shared cognitive architecture matter and can ground predictions about the types of scientific perspectives that we ought to have given our scientific goals and experiences. Moreover, as noted above, different scientific goals can lead to substantively different (normative) perspectives, along with the very real possibility of non-unifiability of the corresponding scientific theories. That is, this perspectivism can potentially lead to pluralism, though the details matter in terms of predicting whether and when pluralism might arise.[5] More generally, this perspectivism implies that "mirror realism" should not necessarily be correct in many cases, but rather we should expect—particularly for sciences that are more focused on measuring and controlling rather than explaining—to find theories that turn out to have various (defensible) misconceptions or falsehoods. There are thus multiple ways in which this type of perspectivism makes substantive claims (that could potentially have turned out to be wrong).

The careful reader will have noticed that nothing I said in the preceding few paragraphs was actually specific to scientific learning and theorizing. Exactly the same points could be made about everyday learning and theorizing. The perspectivism that I defend here results naturally for almost any cognitive agent that must learn about its world and then reason to try to achieve particular goals. For example, our "theories" about the spatial environments in which we move ought, on this account, to be expected to be perspectival in various ways in light of the goals we typically have when navigating those environments (Maguire et al. 2000; Maguire, Woollett, and Spiers 2006). More generally, I contend that we should embrace the type of perspectivism that I defend here, partly because we are all already (or should be) perspectivists about our engagement with the everyday world. Our perspectivism about everyday experiences is (or should be) similarly safe-and-substantive: we are not forced into strong relativism or skepticism about the world, since the world "gets a say" in our perceptions; but we are also not left with vacuous claims about our "contributions" to our understanding of the world.

In this regard, this perspectivism fits closely with the type of view advanced by Chirimuuta (2016). She argues that advocates of scientific perspectivism should base their metaphors and analogies on haptic perception, or perception by touch, rather than visual perception. Haptic perception is clearly mediated by the particular sense organs, rather than purporting to give a "mirror" (perhaps with a subset filter) of the world. We are not under any illusions that our touch-based understanding of the world provides some kind of direct access. Moreover, haptic perception is clearly action-driven: our touch perception is intimately connected with our abilities to influence, move, and manipulate objects in our environment. That is, Chirimuuta's (2016) argument depends on perspectives having exactly the same components that I have discussed here—concepts (so no mirroring) and goals (so actions). More generally, our arguments share the high-level idea that scientific perspectivism is a special case of the perspectivism that arises in our everyday lives. And just like our everyday perspectivism, our scientific perspectivism is as safe-and-substantive as our views about people, penguins, and puppies.

Acknowledgments

Thanks to audiences at the 2016 "Perspectival Modeling: Pluralism & Integration" conference at the University of Edinburgh and at the Center for Philosophy of Science for insightful comments on earlier versions of these arguments. Thanks especially to Mazviita Chirimuuta, Michela Massimi, Sandra D. Mitchell, Joel Smith, and Ken Waters.

Notes

1. Where appropriate, I will note places that this methodological individualism is potentially limiting or distorting.
2. For the purposes of this chapter, I will not worry about the distinctions and relationships between concepts and categories.
3. Importantly, this generalization only holds for models that do not incorporate a decision-theoretic action component into the learning system.
4. Note that my assumption of methodological individualism is doing substantive work here. To the extent that we want to talk about the perspective of a community, then we plausibly have to include external factors of the sort that are often lumped together under terms like "paradigm" or "research program."
5. It is also unclear whether this "non-unifiability" is problematic, at least if we adopt a thoroughgoing goal-based perspectivism about scientific theories. I have elsewhere (Danks forthcoming) argued that the pragmatic perspectivist will almost always have exactly as much unifiability as she wants or needs, even if that falls short of the realist's demands.

References

Chirimuuta, M. 2016. "Vision, Perspectivism, and Haptic Realism." *Philosophy of Science* 83(5): 746–756.

Cohen, J. 2015. "Perceptual Representation, Veridicality, and the Interface Theory of Perception." *Psychonomic Bulletin & Review* 22(6): 1512–1518.

Cohen, M. A., Dennett, D. C., and Kanwisher, N. 2016. "What Is the Bandwidth of Perceptual Experience?" *Trends in Cognitive Science* 20(5): 324–335.

Conway, M. A., and Pleydell-Pearce, C. W. 2000. "The Construction of Autobiographical Memories in the Self-Memory System." *Psychological Review* 107(2): 261–288.

Crane, T. 1992. "The Nonconceptual Content of Experience." In *The Contents of Experience*, edited by Crane, T., 136–157. Cambridge: Cambridge University Press.

Danks, D. 2014. *Unifying the Mind: Cognitive Representations as Graphical Models*. Cambridge, MA: MIT Press.

Danks, D. 2015. "Goal-Dependence in (Scientific) Ontology." *Synthese* 192(11): 3601–3616.

Danks, D. forthcoming. "Unifiability of Pragmatic Theories." In *The Pragmatist Challenge*, edited by Andersen, H. K. Oxford: Oxford University Press.

Denison, S., Bonawitz, E. B., Gopnik, A., and Griffiths, T. L. 2013. "Rational Variability in Children's Causal Inferences: The Sampling Hypothesis." *Cognition* 126(2): 285–300.

Dretske, F. 1981. *Knowledge and the Flow of Information*. Cambridge, MA: MIT Press.

Eimas, P. D., Siqueland, E. R., Jusczyk, P., and Vigorito, J. 1971. "Speech Perception in Infants." *Science* 171(3968): 303–306.

Etcoff, N. L., and Magee, J. J. 1992. "Categorical Perception of Facial Expressions." *Cognition* 44(3): 227–240.

Evans, G. 1982. *The Varieties of Reference*. Oxford: Oxford University Press.

Feltovich, P. J., Spiro, R. J., and Coulson, R. L. 1989. "The Nature of Conceptual Understanding in Biomedicine: The Deep Structure of Complex Ideas and the Development of Misconceptions." In *Cognitive Science in Medicine: Biomedical Modeling*, edited by Evans, D. A., and Patel, V. L., 113–172. Cambridge, MA: MIT Press.

Feyerabend, P. 1975. *Against Method*. London: Verso.

Goldstone, R. L. 1994. "Influences of Categorization on Perceptual Discrimination." *Journal of Experimental Psychology: General* 123(2): 178–200.

Goldstone, R. L., and Hendrickson, A. T. 2010. "Categorical Perception." *WIREs Cognitive Science* 1(1): 69–78.

Hagmayer, Y., Meder, B., Osman, M., Mangold, S., and Lagnado, D. A. 2010. "Spontaneous Causal Learning While Controlling a Dynamic System." *The Open Psychology Journal* 3: 145–162.

Hanson, N. R. 1958. *Patterns of Discovery: An Inquiry Into the Conceptual Foundations of Science*. Cambridge: Cambridge University Press.

Harnad, S. R., ed. 1987. *Categorical Perception: The Groundwork of Cognition*. Cambridge: Cambridge University Press.

Hoffman, D. D., Singh, M., and Prakash, C. 2015. "The Interface Theory of Perception." *Psychonomic Bulletin & Review* 22(6): 1480–1506.

Kitcher, P. 1993. *The Advancement of Science: Science Without Legend, Objectivity Without Illusions*. Oxford: Oxford University Press.

Koutstaal, W., and Schacter, D. L. 1997. "Gist-Based False Recognition of Pictures in Older and Younger Adults." *Journal of Memory and Language* 37(4): 555–583.

Kuhn, T. S. 1962. *The Structure of Scientific Revolutions*. Chicago: University of Chicago Press.

Livingston, K. R., Andrews, J. K., and Harnad, S. 1998. "Categorical Perception Effects Induced by Category Learning." *Journal of Experimental Psychology: Learning, Memory, and Cognition* 24(3): 732–753.

Longino, H. 1990. *Science as Social Knowledge: Values and Objectivity in Scientific Inquiry*. Princeton: Princeton University Press.

Maguire, E. A., Gadian, D. G., Johnsrude, I. S., Good, C. D., Ashburner, J., Frackowiak, R.S.J., and Frith, C. D. 2000. "Navigation-Related Structural Change in the Hippocampi of Taxi Drivers." *Proceedings of the National Academy of Sciences* 97(8): 4398–4403.

Maguire, E. A., Woollett, K., and Spiers, H. J. 2006. "London Taxi Drivers and Bus Drivers: A Structural MRI and Neuropsychological Analysis." *Hippocampus* 16(12): 1091–1101.

Markman, A. B., and Ross, B. H. 2003. "Category Use and Category Learning." *Psychological Bulletin* 129(4): 592–613.

Medin, D. L., Lynch, E. B., Coley, J. D., and Atran, S. 1997. "Categorization and Reasoning Among Tree Experts: Do All Roads Lead to Rome?" *Cognitive Psychology* 32(1): 49–96.

Nichols, W., and Danks, D. 2007. "Decision Making Using Learned Causal Structures." In *Proceedings of the 29th Annual Meeting of the Cognitive Science Society*, edited by McNamara, D. S., and Trafton, J. G., 1343–1348. Austin: Cognitive Science Society.

O'Connor, C. 2014. "Evolving Perceptual Categories." *Philosophy of Science* 81(5): 840–851.

Peacocke, C. 1992. *A Study of Concepts*. Cambridge, MA: MIT Press.

Ross, B. H. 1997. "The Use of Categories Affects Classification." *Journal of Memory and Language* 37(2): 240–267.

Ross, B. H. 1999. "Postclassification Category Use: The Effects of Learning to Use Categories After Learning to Classify." *Journal of Experimental Psychology: Learning, Memory, and Cognition* 25(3): 743–757.

Ross, B. H. 2000. "The Effects of Category Use on Learned Categories." *Memory and Cognition* 28(1): 51–63.

Schacter, D. L., Norman, K. A., and Koutstaal, W. 1998. "The Cognitive Neuroscience of Constructive Memory." *Annual Review of Psychology* 49: 289–318.

Strange, W., and Dittmann, S. 1984. "Effects of Discrimination Training on the Perception of /r-l/ by Japanese Adults Learning English." *Perception & Psychophysics* 36(2): 131–145.

Vul, E., Goodman, N., Griffiths, T. L., and Tenenbaum, J. B. 2014. "One and Done? Optimal Decisions From Very Few Samples." *Cognitive Science* 38(4): 599–637.

Wellen, S. 2015. *The Influence of Practical Goals on Learning: A Theoretical and Empirical Study*. Ph.D. diss., Carnegie Mellon University.

Wellen, S., and Danks, D. 2014. "Learning With a Purpose: The Influence of Goals." In *Proceedings of the 36th Annual Conference of the Cognitive Science Society*, edited by Bello, P., Guarini, M., McShane, M., and Scassellati, B., 1766–1771. Austin: Cognitive Science Society.

Wellen, S., and Danks, D. 2016. "Adaptively Rational Learning." *Minds & Machines* 26(1): 87–102.

8 Charting the Heraclitean Brain

Perspectivism and Simplification in Models of the Motor Cortex

Mazviita Chirimuuta

1 Perspectivism and the Demands of Simplification

It is a frequently stated fact that the human brain is the most complicated object in the known universe; yet it is unclear whether or not this "fact" is a by-product of human vanity or hype drummed up by neuroscientists. To the neurologist Kurt Goldstein, it was not obvious that an invertebrate is more simple than a man, and for that reason he saw no obstacle to selecting the human being as his model organism (Goldstein 1939, 2).[1]

The point is that, since everything in nature is in its own way complicated, claims for complexity need to be made specific. Here I will argue that an important respect in which the brain is complex is that it is *Heraclitean*. By this I mean that the brain, like the Heraclitean river, is the kind of thing that can only maintain its identity by undergoing continual change. John Dupré (2012) has argued that all living organisms should be characterized in this way, that is, as processes rather than entities. In addition, Peter Godfrey-Smith (2016) proposes that the Heraclitean nature of biological cells has important implications for how we understand cognition: it is the important difference between the nervous system and man-made computers.

The fact that the brain is made of living tissue such as neurons (an electrically excitable type of biological cell) means that its constitution is constantly changing with metabolism. As Marder and Goaillard (2006, 563) describe it, "each neuron is constantly rebuilding itself from its constituent proteins, using all of the molecular and biochemical machinery of the cell." Godfrey-Smith's idea is that the inherent changeability of biological tissue was leveraged during the evolution of the nervous system as a means for learning and coping with the challenge of staying alive in an unstable world, and is an important factor in making biological "computation" what it is. This, he argues, puts limits on the functional equivalence between brains and artificial computers.

Looking outside the cranium, John Haugeland (1996) has argued that the fact that the brain is densely interconnected with a mobile body, itself operating in an ever-changing environment, means that the mappings between "inner" neural states and "external" consequences may be constantly evolving. This is more contentious than the biological argument because, as we will see, the empirical evidence for the presence or lack of stable mappings is not conclusive.

What is uncontroversial, however, is that science thrives when complex things can be made to seem simple. Various authors have made the case that complex systems, especially those studied in the biological and behavioral sciences, afford modeling from a variety of perspectives because no one set of theoretical or experimental practices gives the scientist access to all of the relevant phenomena in the domain of interest.[2] Here I build on this work by emphasizing not only the way that scientific perspectives "passively" filter out details not relevant to their own theory and practice but also the way that they "actively" impose simplifying assumptions onto the target system.[3]

My case study in section 2 presents the ongoing controversy between two perspectives on the motor cortex as amounting to a difference in choice of methods selected to cope with the brain's Heraclitean nature. The assumptions embedded in each modeling perspective are effective ways to simplify the brain, but they lead to views about the function of the motor cortex that are apparently contradictory. In section 3 I discuss the philosophical implications of this clash of perspectives. Does it follow that there are non-perspectival truths about the motor cortex (and by extension, the primate brain) that are unknowable to science? In section 4, I voice some support for this Kantian conclusion by presenting a general framework for thinking about the operation of abstraction in science.

2 Two Perspectives on the Motor Cortex

Since everything in nature is complicated in its own way, it is important to recognize simplicity for what it is: something manufactured by means of the scientific method, both materially (in the setting up of laboratory condition) and conceptually (by devising abstract and idealized mathematical representations; Cartwright 1999). Take the proverbial Heraclitean flux: "on those stepping into rivers staying the same other and other waters flow" (Graham 2015). The scientist dipping her toe in the waters finds, indeed, that in a global sense the river stays the same and there is a regularity to its undulations; yet, the constituents of the river in terms of which she would wish to explain its global properties and regularities are themselves ever changing. This poses a challenge which can be met with various strategies, two of which I will discuss here because they are analogs to the two perspectives on the motor cortex that form my case studies.

The first strategy is to approximate to stasis. One constructs a model of the river at a snapshot in time, ignoring its dynamics and changes in composition. The other strategy is to find simple flow patterns. In this case the dynamics are the target of the model, but rather than attempting to represent every tiny current and eddy, one seeks a compact representation of only the major currents that may be repeatedly observed (such as the one parallel to the bank) and any due to large features (such as islets in the stream).

Likewise, when neuroscientists take on the challenge of modeling the Heraclitean brain, a popular strategy is to assume that the response properties of neurons are approximately fixed—that the neural system is one with stable input-output relationships, which can be represented as a mathematical function. In essence, the proposal is that each neuron represents or codes for some state of affairs in the extra-cranial world. This is the *intentional perspective*. An alternative is to model the dynamical evolution of the neural system but to seek a relatively simple set of equations governing it. This is the *dynamical perspective*, and it often (but not always) comes with the denial that neurons code for or represent anything. The neuroscience of the motor cortex is a particularly apt topic here, because there has been a long-standing and often heated dispute over what the function of this brain area is, leading the scientists themselves to be explicit in stating and arguing for their different theoretical perspectives.[4]

2.1 The Intentional[5] Perspective

Near the start of his lecture arguing that the basis of the success of the exact sciences is their ability to find "economical" representations of phenomena, Ernst Mach (1895, 186) tells us

> "Life understands not death, nor death life." . . . Yet in his unceasing desire to diminish the boundaries of the incomprehensible, man has always been engaged in attempts to understand death by life and life by death.

With these words, Mach foresaw why the information-processing and intentional approach would be such a dominant force in the neuroscience of the future. The point is that even though the living brain is inherently Heraclitean, treating it as functionally equivalent to a non-living symbolic system (a computer) has been an effective way to abstract away from the fluid details of neural hardware.

Some of the founders of computational neuroscience, such as Rashevsky, McCulloch, and Pitts, were quite self-aware about the purpose of this kind of abstraction and idealization (Abraham 2002).[6] On the other hand, neuroscientists and philosophers under the influence of functionalist theories of mind have had more of a tendency to interpret the brain-computer analogy in a literal way: to treat even rough functional equivalence as an indicator of sameness at a higher level of description, namely, at the level of the coding scheme or algorithm that both systems are said to implement.

The intentional perspective on the motor cortex is comparable to the "fixed filters" model of the visual system.[7] The core assumption is that individual neurons represent or code some parameters relevant to movements in specific body parts. These may be individual muscle activations, sequences of muscle activations, or higher-order parameters such as the velocity of an arm movement. For example, the cosine tuning model of (Georgopoulos, Schwartz, and Kettner 1986) treats each motor cortical neuron controlling arm movement as firing maximally at its preferred direction of movement, with firing rate dropping away as a cosine function for non-preferred directions. One of the major difficulties for this perspective has been that neurophysiological recordings have yielded partial evidence for each of these hypotheses (and more) about what the motor cortex codes, leading to a lack of consensus within the intentional camp (Omrani, Kaufman, Hatsopoulos, and Cheney 2017). Within the intentional tradition, trial-to-trial variability in neuronal responses is classified as *noise* rather than as variance to be modeled and explained. This is in part for practical reasons (see section 2.2) and in part because a common assumption is that the neurons' tuning properties are fixed, and so variability in responses is not coding anything.

The assumption that motor cortical neurons code for intended movements has found a practical application in brain-computer interface (BCI) technologies that record from this brain region and employ decoding algorithms on the data to derive signals for controlling a robot or cursor. However, it does not follow from the fact that neural data can be decoded in these experiments that the intentional models are realistic or even approximately true of the brain. Certain assumptions made by the decoding algorithms have been shown to be false with respect to neurophysiology of the motor cortex, but during the experiments the brain adapts to biases introduced by the models (Koyama et al. 2010).

2.2 The Dynamical Perspective

The dynamical perspective is the more recent arrival, though its advocates credit Thomas Graham Brown (1882–1965), an associate of Charles Sherrington, with having anticipated their central claim. Speaking of the spinal cord, Brown (1914, 40) writes, "the fundamental activity of this system is the rhythmic." On the current view we are told that the motor cortex is a "pattern-generation machine" (Kaufman, quoted in Omrani et al. 2017, 1835).

One way to summarize the difference between the dynamical and the informational perspective is to say that the relationship of *causation* (between neurons and bodily movements) replaces the intentional relation. While all agree that motor cortical activity is causally upstream of movement, proponents of the dynamical view do not give this an intentional spin (namely, positing that causal interactions between neurons and muscles are merely the medium of information transmission). Instead, they treat the cortex and muscles as coupled oscillatory systems and ask how the cortex orchestrates its sequence of oscillations (of neural population firing) such that they eventually cause an intended sequence of muscle contractions. A basic intuition here is that the oscillations in populations of cortical neurons, at different frequencies and phases, are analogous to a Fourier basis set of sine waves, with which any irregular waveform can be approximated. Likewise, firing patterns in the motor cortex constitute a basis set that, when appropriately deployed, leads to the execution of the range of bodily movements.

Whereas the empirical support for the intentional perspective comes in the form of single neuron tuning curves for movement parameters, the dynamical view has relied on neural population data, processed to show low-dimensional structure. These kinds of data analyses have become common elsewhere in neuroscience with the increase in number of neurons simultaneously recorded; they are the characteristic methods of abstraction within the dynamical perspective.[8] If 100 neurons are recorded during one experimental trial (e.g., an arm reach), the resulting dataset has 100 dimensions (one neuron per dimension). But given the correlations between individual neurons, dimensionality reduction techniques such as principle components analysis (PCA) or factor analysis can typically fit the data into a c.10 dimensional space. The dynamics are represented by plotting the activity of the neural population as a trajectory through a low-dimensional state space.

One feature of the dynamical perspective is that single neurons lose their privileged status when neuroscientists set about trying to interpret cortical function. It is a prediction of this approach that the firing patterns of many of the individual neurons will not be interpretable in terms of external parameters (Kaufman, quoted in Omrani et al. 2017, 1835). Furthermore, Cunningham and Yu (2014, 1501) make the important point that the shift to simultaneous population recordings makes it possible to investigate the causes of trial-to-trial variability, a component of the data that in single-neuron studies is bracketed as noise and dealt with by averaging across multiple trials. This is unavoidable due to the lack of the statistical power in single neuron data that would be needed to support any conclusions as to the source of variability.

Given that the alliance of population data and dynamical modeling[9] has the potential to embrace and explain (rather than average away) some of the trial-to-trial variability in the brain's responses, it may well seem that this perspective takes us closer to the truth of the Heraclitean brain. Indeed, Haugeland (1996, 123), in his recounting of Hubert Dreyfus's challenge to the entire information-processing framework, tells us that a scenario in which there are no stable mappings between brain states and motor outputs (such as letters typed on a keyboard) would completely undermine the notion of a neural code. The resulting picture of embodied, embedded intelligence is one that has been promoted by some practitioners of dynamical modeling in cognitive science.[10] However, it would be far too quick to argue from the existence of trial-to-trial variability in data recorded during repeated movements to the conclusion that there are no roughly stable mappings between patterns of motor cortex activity and resulting bodily movements, that the "content" of the activity patterns is entirely context dependent. For one thing, there are many neurons for which the mappings are reliable enough so that averaging across trials reveals a preference for a particular direction of movement; this is the core result that undergirds the intentional perspective. Also, the cognitive and body context is not the only source of variability; neurophysiological recording techniques are also noisy, and so it is an open question how much trial-to-trial variability is due to behavioral context or due to the recording methods.

Now, a defender of Haugeland or Dreyfus may reply that if it were possible to observe the motor cortex during naturalistic movement conditions, where cognitive and bodily context is uncontrolled—where attention wanders freely, and the posture of the rest of the body is not constrained by harnesses (as happens during experiments on arm-reaching in monkeys)—then the roughly stable mappings would dissipate and be seen for what they are: an artifact of laboratory conditions. This is an interesting conjecture, because if the way that motor cortex activity maps to movement in naturalistic conditions is genuinely Heraclitean, while stability is generated by the constraints of laboratory conditions, then it turns out that the dynamical perspective is not much better placed to represent the Heraclitean motor cortex, in all of its changeable glory, than the intentional one.

The reasons are as follows. First, one aim of the population analyses performed by the dynamical camp is to identify reliable correlations between population (as opposed to single neuron) activity and movements. Even though these correlations are interpreted causally rather than intentionally, and such mappings do not require that *individual* neurons behave in the same way on each trial, so long as some global pattern of activity is maintained (e.g., a certain number of neurons oscillating in a particular way), the research program would be a non-starter if there were absolutely no consistent relationships between movements and neural activity, at any level of description. (This is perfectly consistent with the point made above, that given the statistical power afforded by multineuron simultaneous recording, this approach makes it possible to explain some trial-to-trial variation in terms of behavioral or cognitive context.)

Second, just as the core findings of the intentional perspective (i.e., consistent neuron-movement mappings) may be dependent on the fact that the neural responses are generated in controlled laboratory conditions, the core findings of the dynamical perspective, that the population data can be represented in a low-dimensional state space, which yield hypotheses about the relationship between global state and movement, may themselves be dependent on the same simplifications introduced in the laboratory. Surya Ganguli and colleagues have presented some formal results that, they argue, show that the low-dimensional structure revealed in neural population studies so far is due to the simplicity of tasks used in experiments (Gao and Ganguli 2015; Gao et al. 2017; but see Golub et al. 2018). The upshot is that even if the brain outside of the laboratory is truly Heraclitean in the way that Haugeland and Dreyfus propose, the very techniques used by neuroscientists, in order to observe its workings, tend to make its behavior less complex than this. The Heraclitean brain, one might say, is not an observable object of science.

3 Relating the Perspectives

Table 8.1 summarizes the main points of difference between the two perspectives. These concern kinds of experimental protocol and data analysis as well as the original sources (beyond neuroscience) of the theoretical frameworks. The final line of the table states the two divergent claims about the function of the motor cortex that are made on the basis of these avenues of research: either that this brain area is specialized for coding movements or that it is a pattern generator.

One way to understand the lesson of perspectivism is that the picture of the world offered by science is like a cubist painting: a handful of different

Table 8.1 Comparison of Intentional and Dynamical Perspectives

Intentional Perspective	*Dynamical perspective*
Single neuron physiology	Simultaneous population recording
Represent responses with tuning curve	Represent responses with state space trajectory
Intentional relationship between neurons and movement	*Causal* relationship between neurons and movement[b]
Computer science origin	Physics origin[c]
Formalism in information theory	Formalism in dynamical systems theory
Simplification via abstraction to coding scheme	Simplification via dimensionality reduction
Inherently atemporal[a]	Inherently temporal
Motor cortex *codes movements*	Motor cortex is a *pattern generator*

[a] See Beer and Williams (2015, 5). They present a modified version of information theory that encompasses such temporal changes.

[b] NB: in themselves the data only indicate that there are correlations between neuronal activity and movements. These relationships are then interpreted as intentional or causal ones.

[c] Fairhall (2014, x).

points of view are employed at once, and reality comes through to us only as fragmented and distorted. We have no access to a God's-eye view of nature, and science is not a clear, accurate depiction of reality. One might conclude, therefore, that from the dynamical perspective the motor cortex is a pattern generator, and from the intentional one it is a system for coding movements, and that is the end of the story.[11]

However, many would be dissatisfied if matters were left just there. There are unanswered questions about the relationship between the perspectives: Are they competitors or complementary to one another? Could one perspective ultimately subsume the other? Furthermore, an objection to the perspectivist conclusion, as just stated, is that it is just a version of relativism. Thus, it seems, we are owed a more rigorous notion of perspectival truth in order to avoid this outcome. In section 3.2, I will bring some options to the table here. Before that, I will examine the relationship between the two perspectives.

3.1 Rivals or Allies?

A theme of Sandra Mitchell's (2003) account of integrative pluralism in science is that different perspectives are often complementary to one another, with one view compensating for the deficiencies of another, and cooperation across perspectives occurs when there is a practical challenge that cannot be addressed with one approach alone. Integration does not entail subsumption of one perspective by another; the various models, methods, and representations that constitute a perspective will retain their distinct identities.[12] In keeping with the claims of integrative pluralism, neuroscientist Adrienne Fairhall has argued that an integration between dynamical and informational approaches is a crucial step in the progress of neuroscience:

> Ultimately, the development of methods to map the dynamics of the physical substrate onto the computational is the bottleneck in our ability to truly comprehend the biological mechanisms of intelligence.
> (Fairhall 2014, xi)

A condition of integrative pluralism is that there are pros and cons associated with each perspective, and for that reason they mutually support one another. This raises the question of what the strengths and weaknesses of each of the motor cortex perspectives are. I argued above that the dynamical approach cannot be claimed to give us the unvarnished truth about the Heraclitean brain because of its reliance on finding low-dimensional structures. That said, it is more faithful to the ever-flowing nature of brain processes than the intentional perspective, with its static way of conceptualizing neural functions. So why retain the intentional perspective now that the dynamical one has come of age?

There is something important about the brain that the intentional perspective captures but which eludes the dynamical one. This is the fact that brain states really do seem to be directed to (not merely correlated with or caused by) external states of affairs. These states persist in memory, absent the external stimulus, and are robust to perturbations in the brain. For instance, BCI experiments have shown that if the mappings between neural activity and motor output are perturbed, the motor cortex will adjust its activity patterns to achieve a new set of stable mappings (Jarosiewicz et al. 2008). There is no obvious way to describe the apparent purposefulness of this reorganization within the dynamical framework. More generally, directed relationships between neural states, past events, goals, and future expectations do seem to be critical to explaining what makes a system intelligent, hence they have often been taken to be a "mark of the cognitive."

The dynamical perspective takes the same correlations between neural activity and external states of affairs that undergird the intentional account but lends them no more than causal significance. So it is open to a proponent of the intentional framework to *label* the neural population patterns found using the dynamical methods as *representations of movements*.[13] In the motor cortex case, there is nothing to prevent an intentionalist doing this, but there are not the grounds to insist on this relabeling either.

In the debates that have gone on between proponents of the two perspectives on the motor cortex, the tone has not been particularly conciliatory. The background assumption, it seems, is that the two approaches are natural rivals and not allies—that ultimately one view must be right and the other wrong. In particular, advocates of each side have taken pains to show that the core phenomena stated by their opponents are recoverable within their own modeling framework (e.g., Kaufman in Omrani et al. 2017). The title of a comparison study by Michaels, Dann, and Scherberger (2016) is telling: "Neural Population Dynamics During Reaching Are Better Explained by a Dynamical System Than Representational Tuning."

In contrast, one prominent figure in the field of dynamical modeling, Randall Beer, has made the case for an alliance between dynamical and informational approaches. Beer and Williams (2015) show that the behavior of one extremely simple, simulated cognitive agent can be explained either using the formalism of information theory (IT) or dynamical systems theory (DST). They write:

> As mathematical theories, IT and DST can be applied to any system that takes the proper form to meet their defining requirements; they intrinsically make no scientific claim as to "what's really going on." Instead, they are best viewed as distinct mathematical lenses through which we can examine the operation of a system of interest.
>
> (Beer and Williams 2015, 2)

These authors go on to say that "the mathematical languages themselves are merely more or less useful to a given purpose" (Beer and Williams 2015, 23). Inspired by these remarks, one might attempt a reconciliation between the two perspectives via an instrumentalism which denies that mathematical models afford any view of underlying nature beyond the empirical predictions. As such, the use of these mathematical tools tells us nothing about what the brain is really like.

3.2 On Perspectival Truth

Embracing this conclusion, for the perspectives on the motor cortex, would invite the objection that perspectivism is nothing more than instrumentalism rebranded (Morrison 2011). Michela Massimi is one philosopher of science who has taken pains to show that perspectivism is actually a version of scientific realism, by developing a substantial and non-relativistic notion of perspectival truth. In this section I will examine whether or not her account is applicable to the motor cortex example.

One version of perspectival truth that Massimi (2018, 349) considers is a contextualist one:

(P$_3$) *Perspective-dependence$_3$*. Knowledge claims in science are perspective-dependent$_3$ when their truth-conditions depend on the scientific perspective in which such claims are made.

Here, scientific perspectives provide the context in which truth conditions are defined for the knowledge claims of science. For example, on this account

<M1 neurons code movement parameters>

would be true in the context of the intentional/informational perspective and false in the dynamical one. The stated benefits of P$_3$ are that it upholds the realist intuition that science gets things (partially) right but at the same time rejects a monistic view of scientific knowledge in favor of plurality of perspectives that offer "idealized, inaccurate, and yet still true perspectival images of an independent world" (Massimi 2018, 353). The downside, Massimi contends, is that if P$_3$ is all we aim for, we must also concede that nature-in-itself is an unknowable, noumenal reality. Massimi's more ambitious notion of perspectival truth employs a distinction between context-of-use and context-of-assessment:

(P$_4$) *Perspective-dependence$_4$*. Knowledge claims in science are perspective-dependent$_4$ when their truth-conditions . . . depend on the scientific perspective in which such claims are made. Yet such knowledge claims must also be assessable from the point of view of other . . . scientific perspectives.

(Massimi 2018, 254)

This idea is fleshed out with the example of the claim

<Water is a liquid with viscosity>

which is true from the perspective of hydrodynamics but false according to statistical mechanics. However, if statistical mechanics is deployed as a "context of assessment" for hydrodynamics, it can be shown that the property of viscosity is still recoverable in statistical mechanics "as a derivative property" (Massimi 2018, 354). Thus, the knowledge claim of one perspective is validated by the other perspective after all.

There is a parallel to Massimi's account of viscosity in the motor cortex case: proponents of each perspective do claim to be able to recover the core phenomena of the alternative perspective using their own models and assumptions. That is, advocates of the dynamical view have emphasized that the correlations between movements and neural activity, which are taken by intentionalists to be the signature of coding, are also predicted by the dynamical account. Likewise, neuroscientists defending the intentionalist framework hastened to show that curved trajectories in the low-dimensional jPC space, argued by Churchland et al. (2012) to be evidence for the dynamical view, were consistent with cosine tuning in the motor cortex (see also Michaels et al. 2016). Yet the conclusion that each

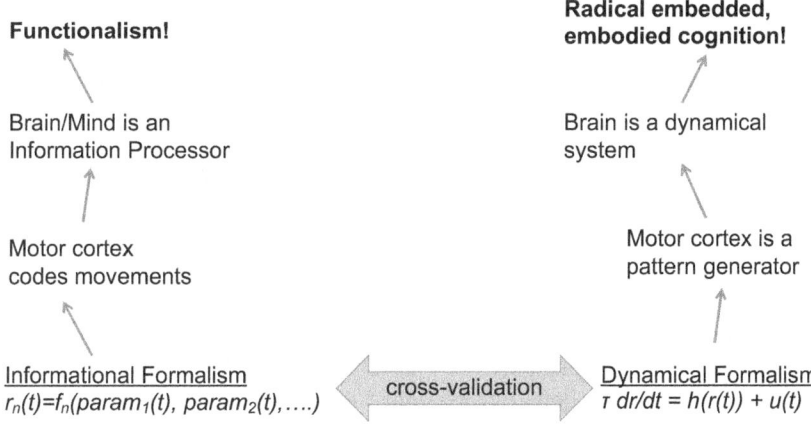

Figure 8.1 Illustration of cross-validation of quantitative models and divergence of qualitative interpretations of those models. Beer and Williams (2015) provide a demonstration of cross-perspective validation for their very simple, minimally cognitive agent.

side draws from any instances of cross-validation between perspectives is not the conciliatory or pluralist one; rather, such findings are claimed to *undercut* the other perspective. The logic is that if an alternative perspective is not needed to explain a portion of the observed findings, then only one perspective should be employed.

I should emphasize that there may well be sociological and psychological reasons why the debate over motor cortex perspectives has more of the look of a turf war than a display of polite recognition of the need for diversity of theories and methods in science. But one philosophical explanation for why diversity is not met with tolerance here is that there is no big picture of neural and cognitive function that the different perspectives are converging on. From each individual perspective you get an "interpretable" picture of what the motor cortex fundamentally is, but when placed together, with the pluralist claim that both are in some sense true, the picture becomes incoherent.

The situation is illustrated in Figure 8.1. At the level of mathematical formalism, the intentional and DST perspectives do cross-validate one another in the way that Massimi requires for her robust notion of perspectival truth. As Beer and Williams (2015) show for their minimal cognitive agent, some of the phenomena isolated by dynamical modeling also show up in the informational model and vice versa. However, when one moves beyond the pure formalism, to the level of interpretation of the models,[14] the perspectives diverge. Figure 8.1 presents additional layers of interpretation, which end ultimately with different philosophical views about what the brain/mind fundamentally is.

Each perspective provides mathematical formalisms for describing neural activity. At this purely quantitative level, the perspectives can be shown to be consistent with one another, satisfying P_4. When one considers the qualitative descriptions of the neural systems associated with each perspective (the claims about what the cortex "is like"), over and above the mathematical formalism, inconsistency appears. Note that *functionalism* is a philosophical theory of the mind-body relationship which often presupposes internal representations; in contrast, the *embodied-embedded cognition* theory denies that representations are needed in order to explain mental capacities. I do not suppose that many neuroscientists employing these models also commit themselves to these philosophical theories, but it is relevant that the models do lend themselves to these higher levels of interpretation.

Since the clash of perspectives only comes with the interpretation of the mathematical models, one response to this difficulty is to strongly discourage interpretation. The task of the neuroscientist is to shut up and calculate, and leave speculation about the nature of the brain/mind to the philosophers. This response is in keeping with Beer and William's recommendation that the mathematical theories of IT and DST by themselves make no claims about "what's really going on."

A pluralism that blends into instrumentalism by putting restrictions on extra-empirical interpretation is not an option for Massimi, since a non-negotiable claim of scientific realism is that the successful theories and models employed by scientists can also be interpreted in order to tell us something about the underlying nature of things. Massimi argues that for perspectivism to be made compatible with realism, the different perspectives must be shown to endorse each other's knowledge claims; but in our case the mutual reinforcement is only possible at the level of uninterpreted mathematical formalism. Once interpretation is lent to the models, the clash of perspectives is jarring.

This amounts to a dilemma for the ambitious perspectival realism advocated by Massimi. P_4 requires the knowledge claims of each perspective to be endorsed by the other perspectives. In the IT/DST example the endorsement is granted only if the cross-perspective assessment is restricted to the quantitative predictions of the mathematical models, while no cross-perspective endorsement is granted to the qualitative interpretations associated with those models—the bigger picture they offer as to "what's really going on" in the motor cortex, and the brain more generally. So *either* P_4 knowledge claims must be restricted to the quantitative predictions of the mathematical models, leading to instrumentalism *or*, if not so restricted, it is clear that convergence of knowledge claims across perspectives does not obtain; this implies that one of the perspectives is the truer one and pluralism must be abandoned.[15] Thus, it seems, the ambitious notion of perspectival truth, P_4, is not applicable to our case. I will conclude the chapter with some thoughts on why the more modest P_3 is not such a bad notion to settle for.

4 Conclusion: Two Philosophical Perspectives on Abstraction

It is interesting that the word "abstraction" bears two different meanings in contemporary philosophy: one lofty, the other mundane. In the lofty sense, an abstraction is an abstract entity, not spatially and temporally located and so possibly residing in Plato's heaven. Speaking mundanely (the use more common among philosophers of science), abstraction is synonymous with simplification and, paired with idealization, an important model-building strategy employed by scientists. These two conceptions of abstraction animate two very different explanations for the "unreasonable effectiveness of mathematics"; that is, they provide two different answers to the question of why mathematics is such a useful tool in science. The lofty explanation is that the underlying reality of nature consists in mathematical structure, and the task of the exact sciences is to discover these. The mundane one is that science progresses when humans find ingenious ways to simplify complex phenomena, and mathematics is the preeminent tool for doing this. Adherents to the lofty way of thinking about abstraction and the role of mathematics in science are in good company—not only Plato but Galileo (stating that the book of nature is written in the language of mathematics), Descartes (and other rationalist philosophers), and contemporary ontic structural realists have intellectual kinship here. However, this Platonic tradition meets difficulty with the existence of pluralities of different kinds of mathematical representation of natural phenomena. If the construction of a predictively powerful model of natural phenomena is also, in some sense, a revelation of the mathematical laws that underlie observable phenomena, then how can it be that the book of nature seems to be written by multiple authors?[16]

Kant is a figurehead for the mundane approach. Instead of taking the content of the abstract mathematical representation to be somehow more fundamental than the concrete, observable events, on this account one regards the mathematical representations as a set of structures employed by the human mind as a means to order the observable phenomena.[17] When confronted with the varying, multifaceted, and disordered events in nature, mathematics offers a useful set of structures for imposing representational order on them, especially by means of leaving out details—the process of abstraction. Once one makes the Kantian move of looking "inwards" for the explanation of how mathematical representations yield knowledge of nature, it is not jarring or surprising that there are multiple ways to achieve order and abstract, hence there may be a plurality of kinds of mathematical model for the same piece of nature.

This way of thinking about abstraction and the scientific method was popular in the philosophy of science a century ago, even if neglected now.[18] I believe it offers significant benefits for thinking about perspectival pluralism. Not only does it welcome the existence of multiple perspectives, but it also permits a relaxed response to the possibility of there being unknowable, non-perspectival truths. In our case we can say that the brain-in-itself is not knowable in its full, Heraclitean complexity because no human scientist (with limited, human cognitive powers) would be able to theorize it completely and accurately as such. The brain-in-itself is not mysteriously unthinkable; it is just very complicated. At the same time, the Heraclitean brain provides constraints on what counts as an acceptable representation of it, and this means that contextual knowledge claims about it—as in P_3—are not fictitious or relativistic ones. To see as through a glass darkly is still to see something.[19]

Acknowledgments

I would like to thank audience members at the Philosophy, Psychology and Neuroscience Research Seminar (University of Glasgow); the Philosophy Work in Progress Seminar (University of Birmingham); the Perspectivism Reading Group (University of Pittsburgh); and the Neural Mechanisms Online (University of Turin) seminar series for lively discussions and many useful suggestions. I am also grateful to Mark Churchland for a discussion of his research, and to the editors of this volume for their many helpful comments on the text.

Notes

1. In the case of *Caenorhabditis elegans*, a tiny worm, it seems that Goldstein is vindicated by current neuroscience:

   > a counterintuitive finding in *C. elegans* is that there is no such thing as "simplicity" despite the reduced connectome (302 neurons, 6963 synapses, 890 gap junctions), even at the earliest stage of sensory processing.
   > (Frégnac 2017, 473)

2. See, e.g., Mitchell (2003) and Longino (2013).
3. Philosophers of science typically refer to the former process as "abstraction" and the latter as "idealization." I think that the distinction between the two is less clear-cut than is normally supposed, so the abstractions I discuss below should not be thought of as *pure* simplifications.
4. Omrani et al. (2017) is a very helpful review of the current approaches to motor cortex. There are more options than the two perspectives I present here, but space does not permit discussion of these. See also Scott (2008).
5. I use the word "intentional" instead of "representational" to avoid confusion with the scientific representations (models, maps, etc.) that come up in the discussion of scientific perspectives. Shenoy, Sahani, and Churchland (2013) and Churchland et al. (2012) refer to it as the "representational perspective."
6. Rashevsky, for example, says:

   > Following the fundamental method of physicomathematical sciences, we do not attempt a mathematical description of a concrete cell, in all its complexity. We start with a study of highly idealized systems, which at first may not even have any counterpart in real nature.
   > (quoted in Abraham 2002, 16)

7. That is, the idea that individual neurons in the visual system selectively respond to a particular kind of stimulus and that these tunings are stable across time—independent of task and stimulus context. See Chirimuuta and Gold (2009) for discussion.
8. It bears emphasis that the existence of low-dimensional structure in high-dimensional neural population data is a precondition of the dynamical approach. This is how Cunningham and Yu (2014, 1507) compare the intentional and dynamical perspective in terms of their different strategies for simplifying the brain:

   > One of the major pursuits of science is to explain complex phenomena in simple terms. Systems neuroscience is no exception, and decades of research have attempted to find simplicity at the level of individual neurons. Standard analysis procedures include constructing simple parametric tuning curves and response fields, analyzing only a select subset of the recorded neurons, and creating population averages. . . . Recently, studies have begun to embrace single-neuron heterogeneity and seek simplicity at the level of the population as enabled by dimensionality reduction.

   Cunningham and Yu provide a clear review of the methods described in this paragraph.

9. This is not to imply that dynamical systems theory cannot be applied to single-neuron data, as many studies attest.
10. E.g., van Gelder (1995) and Chemero (2009).
11. Cf. Giere (2006, 5–6): "the strongest claims a scientist can legitimately make are of a qualified, conditional form: 'According to this highly confirmed theory (or reliable instrument), the world seems to be roughly such and such.'"
12. See Mitchell 2003; Mitchell, this volume.
13. I thank David Bain, Fiona MacPherson, and Scott Sturgeon for pressing this point.
14. By "interpretation" here I mean, roughly, what is implied by the model over and above its quantitative predictions; cf. the semantic content that scientific realists—but not instrumentalists—would associate with scientific theories. Interpretation can be thought of as an informal representation of the system. Note that for the scientific realist—but not the instrumentalist—the mathematical formalism can also be taken to provide a representation of the neural system.
15. That is not to imply that anyone knows which perspective is the truer one. Furthermore, it could also be that all of the current perspectives are equally far from the truth.
16. Given Beer and Williams's finding of the compatibility of the two formalisms, someone sympathetic to the Platonic view might respond here that, yes, the brain partakes of these two different formalisms and we should desist from trying to interpret them and reconcile the interpretations. My response would again be that imposing any such ban on interpretation leads you into instrumentalism. As Robert Briscoe and Will Davies have pointed out to me, more problematic cases for my arguments are ones where scientists have endeavored to merge the perspectives both in terms of formalism and interpretation. In the case of dynamical and intentional models of the motor cortex, I have not yet come across work of this kind.
17. NB: my characterization of a Kantian view is not aiming at accurate exegesis of Kant's thought on mathematical science, nor does it come with any commitment to Kantianism regarding the ontology of numbers or epistemology of mathematics. For instance, one could think that mathematics is learned by the mind's apprehension of Platonic Forms but still be a "Kantian" in the sense relevant here, i.e., by denying that mathematical structures are the truer reality underlying the appearances in nature and asserting that the utility of mathematics in science comes from the mind's ability to employ certain simple structures in its apprehension of nature. However, there is a connection between the Platonic tradition I characterize here and Platonism regarding the ontology of numbers, in that the indispensability arguments for the existence of numbers presuppose the lofty explanation for the success of mathematical science. I thank Alastair Wilson for this point.
18. I am thinking here of Duhem (1954), Cassirer (1957), Husserl (1970), and Whitehead (1938). In particular, the pluralism advocated here is more in the spirit of neo-Kantians such as Cassirer than Kant himself. Cassirer (1957, 409) makes a point highly relevant to my study, that the exact sciences are only concerned with events "under the aspect of [their] repeatability." We can say that the Heraclitean brain in its never-exactly-repeating richness is simply not a concern to the mathematical neuroscientist; the brain must be presumed, qua object of mathematical neuroscience, to be non-Heraclitean.
19. We might also think of these as two theological outlooks. On the Platonic side, God is conceived of as a mathematically informed creator, and the structure of his creation is revealed through functions and number. Creation is intelligible to human reason to the extent that we understand mathematics.

On the other side we have a negative theology, where God and his ways are somewhat inscrutable to the finite human mind:

> He [the scientist] will choose a certain formula because it is simpler than the others; the weakness of our minds constrains us to attach great importance to considerations of this sort. There was a time when physicists supposed the intelligence of the Creator to be tainted with the same debility.
>
> (Duhem 1954, 171)

References

Abraham, T. H. 2002. "(Physio)logical Circuits: The Intellectual Origins of the McCulloch-Pitts Neural Networks." *Journal of the History of the Behavioral Sciences* 38(1): 3–25.

Beer, R. D., and Williams, P. L. 2015. "Information Processing and Dynamics in Minimally Cognitive Agents." *Cognitive Science* 39(1): 1–38.

Brown, T. G. 1914. "On the Nature of the Fundamental Activity of the Nervous Centres; Together With an Analysis of the Conditioning of Rhythmic Activity in Progression, and a Theory of the Evolution of Function in the Nervous System." *Journal of Physiology* 48(1): 18–46.

Cartwright, N. D. 1999. *The Dappled World: A Study of the Boundaries of Science*. Cambridge: Cambridge University Press.

Cassirer, E. 1957. *The Philosophy of Symbolic Forms; Volume 3: The Phenomenology of Knowledge*. New Haven: Yale University Press.

Chemero, A. 2009. *Radical Embodied Cognitive Science*. Cambridge, MA: MIT Press.

Chirimuuta, M., and Gold, I. J. 2009. "The Embedded Neuron, the Enactive Field?" In *The Oxford Handbook of Philosophy and Neuroscience*, edited by Bickle, J., 200–225. Oxford: Oxford University Press.

Churchland, M. M., Cunningham, J. P., Kaufman, M. T., Foster, J. D., Nuyujukian, P., Ryu, S. I., and Shenoy, K. V. 2012. "Neural Population Dynamics During Reaching." *Nature* 487: 51–56.

Cunningham, J. P., and Yu, B. M. 2014. "Dimensionality Reduction for Large-scale Neural Recordings." *Nature Neuroscience* 17: 1500–1509.

Duhem, P. 1954. *The Aim and Structure of Physical Theory*. Princeton: Princeton University Press.

Dupré, J. 2012. *Processes of Life*. Oxford: Oxford University Press.

Fairhall, A. 2014. "The Receptive Field Is Dead. Long Live the Receptive Field?" *Current Opinion in Neurobiology* 25: ix–xii.

Frégnac, Y. 2017. "Big Data and the Industrialization of Neuroscience: A Safe Roadmap for Understanding the Brain?" *Science* 358(6362): 470–477.

Gao, P., and Ganguli, S. 2015. "On Simplicity and Complexity in the Brave New World of Large-Scale Neuroscience." *Current Opinion in Neuroscience* 32: 148–155.

Gao, P., Trautmann, E., Yu, B. M., Santhanam, G., Ryu, S. I., Shenoy, K. V., and Ganguli, S. 2017. "A Theory of Multineuronal Dimensionality, Dynamics and Measurement." *bioRxiv* 214262. https://doi.org/10.1101/214262.

Georgopoulos, A., Schwartz, A., and Kettner, R. 1986. "Neuronal Population Coding of Movement Direction." *Science* 233(4771): 1416–1419.

Giere, R. 2006. "Perspectival Pluralism." In *Scientific Pluralism*, edited by Kellert, S. H., Longino, H. E., and Waters, C. K., 26–41. Minneapolis: University of Minnesota Press.

Godfrey-Smith, P. 2016. "Mind, Matter, and Metabolism." *Journal of Philosophy* 63: 481–506.

Goldstein, K. 1939. *The Organism: A Holistic Approach to Biology Derived From Pathological Data in Man*. Salt Lake City: American Book Company.

Golub, M. D., Sadtler, P. T., Oby, E. R., Quick, K. M., Ryu, S. I., Tyler-Kabara, E. C., Batista, A. P., Chase, S. M., and Yu, B. M. 2018. "Learning by Neural Reassociation." *Nature Neuroscience* 21: 607–616.

Graham, D. W. 2015. "Heraclitus." In *The Stanford Encyclopedia of Philosophy*, Fall 2015 edition, edited by Zalta, E. N. https://plato.stanford.edu/archives/fall2015/entries/heraclitus/.

Haugeland, J. 1996. "Body and World: A Review of What Computers Still Can't Do: A Critique of Artificial Reason (Hubert L. Dreyfus)." *Artificial Intelligence* 80: 119–128.

Husserl, E. 1970. *The Crisis of European Sciences and Transcendental Phenomenology*. Evanston: Northwestern University Press.

Jarosiewicz, B., Chasea, S. M., Fraser, G. W., Velliste, M., Kass, R. E., and Schwartz, A. B. 2008. "Functional Network Reorganization During Learning in a Brain-Computer Interface Paradigm." *Proceedings of the National Academy of Sciences* 105(49): 19486–19491.

Koyama, S., Chase, S. M., Whitford, A. S., Velliste, M., Schwartz, A. B., and Kass, R. E. 2010. "Comparison of Brain-Computer Interface Decoding Algorithms in Open-loop and Closed-loop Control." *Journal of Computational Neuroscience* 29(1–2): 73–87.

Longino, H. E. 2013. *Studying Human Behavior*. Chicago: Chicago University Press.

Mach, E. 1895. "On the Economical Nature of Physical Enquiry." In *Popular Scientific Lectures*, translated by McCormack, T. J., 186–213. Chicago: Open Court.

Marder, E., and Goaillard, J. M. 2006. "Variability, Compensation and Homeostasis in Neuron and Network Function." *Nature Reviews Neuroscience* 7: 563–574.

Massimi, M. 2018. "Four Kinds of Perspectival Truth." *Philosophy and Phenomenological Research* 96(2): 342–359.

Michaels, J. A., Dann, B., and Scherberger, H. 2016. "Neural Population Dynamics During Reaching Are Better Explained by a Dynamical System than Representational Tuning." *PLOS Computational Biology* 12(11): e1005175.

Mitchell, S. D. 2003. *Biological Complexity and Integrative Pluralism*. Cambridge: Cambridge University Press.

Morrison, M. 2011. "One Phenomenon, Many Models: Inconsistency and Complementarity." *Studies in History and Philosophy of Science Part A* 42(2): 342–351.

Omrani, M., Kaufman, M. T., Hatsopoulos, N. G., and Cheney, P. D. 2017. "Perspectives on Classical Controversies About the Motor Cortex." *Journal of Neurophysiology* 118(3): 1828–1848.

Scott, S. H. 2008. "Inconvenient Truths About Neural Processing in Primary Motor Cortex." *The Journal of Physiology* 586(5): 1217–1224.

Shenoy, K. V., Sahani, M., and Churchland, M. M. 2013. "Cortical Control of Arm Movements: A Dynamical Systems Perspective." *Annual Review of Neuroscience* 36: 337–359.

van Gelder, T. 1995. "What Might Cognition Be If Not Computation?" *The Journal of Philosophy* 92(7): 345–381.

Whitehead, A. N. 1938. *Science and the Modern World*. New York: Macmillan.

9 Cancer Modeling

The Advantages and Limitations of Multiple Perspectives

Anya Plutynski

This dialectical tension, as it may be called, between a realist and an instrumentalist attitude, existing together without contradiction, seems to me characteristic of the deepest scientists.

(Stein 1989, 64)

1 Introduction

In his *Structure of Scientific Revolutions*, Thomas Kuhn offered up a controversial account of scientific change. According to this picture, competing paradigms in the history of science are incommensurable. Scientific revolutions are revolutionary exactly because they involve a radical shift in worldview, or a "Gestalt" shift. Kuhn claimed that scientists operating under different paradigms quite literally "see" the world differently: "after Copernicus, astronomers lived in a different world" (Kuhn 2012, 117). This picture of scientific change has been enormously influential—particularly in the context of the realism/antirealism debate in the philosophy of science.

Kuhn is typically read as a "constructivist" and an "antirealist."[1] Kuhn resisted the view that the aim of science is "a permanent fixed scientific truth, of which each stage in the development of scientific knowledge is a better exemplar" (Kuhn 2012, 172–173). Instead, he compared scientific change to evolution by natural selection. Just as evolution is not directed at a fixed goal, so too our best scientific theories are at best good enough to survive the trials they face, at least relative to the going alternatives, at this particular time and place. Success in science is not a view from nowhere, but the best possible view from here.

On the one hand, many historically and naturalistically inclined philosophers of science have been sympathetic with Kuhn. Scientists are not in the business of gaining, as Giere puts it, a "view from nowhere or everywhere at once" (Giere 1999, 80), but with solving much more circumscribed problems or addressing very specific questions, in service of which they often generate models that are intended to represent the world in some respects, and to a greater or lesser degree of accuracy (Giere 1999).

Fit of a model to the world is a matter of meeting standards that are to some extent conventional and contingent upon historically available tools. The methodological standards and, indeed, the conceptual frameworks of a given disciplinary specialty are to a large extent socially and historically "situated" (Massimi 2018a, 345). Moreover, one often finds that, as Massimi explains, "both across historical periods, and in any given historical period, science witnesses a plurality of models, theories, experimental techniques, and measurement apparatuses—all designed to investigate the very same target system" (Massimi 2018a, 344). That is, even within a particular period, scientists may deploy a variety of different and sometimes apparently inconsistent models, all in service of investigating the same general phenomena.

How can scientists consistently deploy models making different and apparently incompatible assumptions? One answer to this question is to suggest that any incompatibilities will or should ultimately be jettisoned. Another answer is that we ought to view models at best as tools for prediction. The former view is favored by realists; the latter view might be favored by antirealists or instrumentalists. The debate has now raged for decades. Is there a middle ground in this debate?[2] Giere (1999) proposed a view he called "perspectival realism" as a sort of middle ground between extremes on this continuum:[3]

> Rather than thinking of the world as packaged in sets of objects sharing definite properties, think of it as indefinitely complex, exhibiting many qualities that at least appear to vary continuously. One might then construct maps that depict this world from various perspectives. . . . Here we have a way of combining what is valuable in both constructivism and realism. . . . We can agree that scientific representations are socially constructed, but then we must also agree that some socially constructed representations can be discovered to provide a good picture of aspects of the world.
>
> (Giere 1999, 26)

Giere's proposal has two parts: one metaphysical and another epistemic. The metaphysical thesis is that the world itself is "complex"; or, objects are not fixed and easily demarcated, and their properties often vary continuously.[4] The epistemic claim is that, given the complexity of the world as we find it, our scientific representations of the world are (inevitably) partial, or in some sense "perspectival." Giere's appeal to mapmaking is central to his argument; scientists build models on the same sorts of principles that mapmakers build maps. Like maps, models represent the world only partially, or represent the world only in some aspect, and with some degree of accuracy. Indeed, some aspects of models may be deliberately fictional or involve misrepresentations. Just as maps require interpretation, scientific models must be interpreted in part by appeal to the intentions or interests of the modeler. One needs to know which aspects of the model are intended to represent which parts of the world in order to determine if a model is a good one. Giere thus sums up two main features of perspectivism:

> First, there is no total or universal perspective, or, alternatively, there is no perspective from nowhere or from everywhere at once. All perspectives are partial relative to their objects. Second, there is something real that each perspective is a perspective of.
>
> (Giere 1999, 80)

Giere's perspectivism is thus "realist" in the sense that our scientific perspectives are of the world: we can and should interpret models as representing the world more or less accurately. Yet, like Kuhn, Giere grants that what and how scientists come to know is shaped in part by scientists' historical context and particular interests.

Recently, several philosophers of science have either critiqued or refined Giere's perspectival realism (Chakravartty 2010; Morrison 2011; Chirimuuta 2016; Massimi 2018a). All these authors wish to defend a form of realism; however, they disagree about what to make of the fact that scientists often deploy diverse, sometimes apparently incompatible models of the same systems. For instance, some argue that seemingly incompatible models are, as a matter of fact, not incompatible, because each describe the actual dispositions of a real-world target, but differentially revealed by different methods of detection (Chakravartty 2010). Others contend that while in some cases inconsistent models are simply incompatible, in that they ascribe inconsistent "fundamental properties" to the target of inquiry (Morrison 2011, 351), in other cases apparently inconsistent models are simply different ways of elaborating the same set of basic principles. A unified treatment based on shared basic principles is possible, once the falsely attributed properties are jettisoned. Others still argue that the relationships between models and the world are "haptic" rather than "perspectival": interactive, interested, and historically situated (Chirimuuta 2016). Similarly, some claim that across perspectives successful claims of scientific knowledge from one perspective may also meet standards of adequacy when assessed from other perspectives (Massimi 2018a).

In my view, each of these authors provides insight into scientific models and modeling. The key to reconciling them is to recognize that modelers use models for different purposes. Sometimes unity and assimilation is the goal, for the reasons Chakravartty and Morrison suggest; however, sometimes apparent disunity and inconsistency are not only tolerated but also, indeed, maintained (apparently) indefinitely. In part, this may be because scientists often do not know which of the two is the likely outcome ahead of time! As Stein has argued, the very best scientists straddle the "dialectical tension" between realism and instrumentalism. By way of example, Stein points to Maxwell's treatment of the ether:

> If something is transmitted from one particle to another at a distance, what is its condition after it has left the one particle and before it has reached the other? . . . In fact, whenever energy is transmitted from one body to another in time, there must be a medium or substance in which the energy exists after it leaves one body and before it reaches the other, for energy, as Torricelli remarked, "is a quintessence of so subtle a nature that it cannot be contained in any vessel except the Inmost substance of material things." Hence all these theories lead to the conception of a medium in which the propagation takes place. . . . If we admit this medium as an hypothesis, I think it ought to occupy a prominent place in our investigations, and that we ought to endeavour to construct a mental representation of all the details of its action, and this has been my constant aim in this treatise.
>
> (Maxwell 1881, 438)

Like Maxwell, many scientists seem to be capable of holding in mind simultaneously both realist and instrumentalist stances: holding that a theoretical posit is real "enough" to exhibit regularities that can be investigated, but granting at the same time that their role in one's theory or model may at best be instrumentally useful in inspiring further inquiry. Or so I will argue below.

It is also true, of course, that scientific models are constructed in service of performing different functions. All models are in some sense a representation of the world, but representations may be intentionally sketchy, contain deliberate misrepresentations of some part or feature of a system, or be intended (for now) as simply predictive tools. If we grant this, the realist and antirealist need not part ways. Moreover, while some models are general, many treat only a subclass of cases of phenomena that meet specific (restricted) conditions. In such cases, there is no direct contradiction between models; a unified picture that incorporates all such models is not the ultimate aim. Fagan (2017) has recently drawn upon the literature on modeling and perspectival realism to generate a taxonomy of different sorts of relations that might obtain between models:

- Direct conflict
- Simple additivity
- Subsumption
- Interactive process
- Cross-perspective translation
- No cumulative interaction
- Non-interaction
- Complementarity.

She explains and describes these relations as follows:

> Direct conflict is the objectivist view, such that we select the best model among alternatives. Simple additivity is the bare conjunction of statements from different models associated with different methods. Subsumption by more basic principles (traditional unification) is an indirect relation between models—each is subsumed by the shared basic principles. The idea of an interactive process is not a relation per se, but part of a more general framework for thinking about these relations (see below). Cross-perspective translation is a kind of interactive process, but with distinctive features. . . . Non-interaction is a limit case: absence of a relation between models in practice. . . .
>
> . . . Simple additivity involves no substantive connection between models from different perspectives; their contributions are simply strung together in a conjunction. The "unification" of simple additivity is bare logical consistency. . . .
>
> . . . Complementarity is a familiar relation . . . models from different perspectives can complement one another through differences that do not converge on a common core . . . models in different perspectives are, at some point in their construction, fitted together like jigsaw puzzle pieces. I therefore term this category of relation "complementary." Cross-checking and mutual constraint on possibility space are examples.
>
> (Fagan 2017, 27–28)

This taxonomy will be useful in our consideration of the varieties of model-relation in the context of cancer research. In my view, there is a patchwork of kinds of relation between models in cancer research. Sometimes, apparently competing or inconsistent models of cancer initiation or progression are simply concerned with different questions, or different targets, at different scales of analysis. Ultimately, however, I will argue that complementarity, not conflict, is the proper view of the relationship between purportedly competing perspectives on carcinogenesis that I consider in some detail below.

First, however, it is important to note briefly that there are many kinds of things referred to as "models" in cancer science, some of which are "concrete" and some of which are more "abstract." For instance, model organisms and cancer cells in culture are often treated as "models" or experimental systems for investigating various features or aspects of cancer. In contrast, more abstract models include mathematical representations of cancer's dynamics using ordinary differential equations, agent-based computer simulations of changes in cancer cells over time, network models of signaling pathways, or "box and arrow" diagrams of core causal pathways in the cell associated with tumorigenesis. The aims of these more abstract models are diverse. Sometimes they serve as a starting point for developing and testing causal hypotheses; sometimes they are used to make very specific predictions, suggest avenues for intervention, or simply for pedagogical purposes. Scientists may wish to investigate the conditions on or features of a kind of dynamic process; such investigations are more theoretical in character. However, much of modeling in cancer research is concerned with practical matters, such as predicting the course of a cancer type or subtype, or estimating the age of onset, likely response to chemotherapy, or threat to mortality of different cancers. Some of the models are very general; some are highly specific. Given the variety of functions they are intended to serve, and the variety of things scientists characterize as "models," it would be difficult at best, and foolish at worst, to attempt to offer a general account of the relations between and functions of models in cancer science.

Nonetheless, Giere's picture of the relationship between "theories" and "models" is helpful as a first pass when considering the construction and roles of formal models of cancer. Formal models of cancer's dynamics or progression are in large part informed by or built on the principles of "theories" (or, as I prefer, research traditions),[5] though such principles are often consistent with a variety of different modeling strategies. For instance, some mathematical models built on the principle that cancer is a multistage process driven largely by acquisition of mutations represent cancer initiation and progression as akin to a dynamic, evolving population; others take populations of cells in a tumor as engaged in a process of "competition" akin to competitive exclusion modeling in ecology. Models built on the principle that cancer is shaped by epigenetic factors that affect developmental pathways may represent cancer as a shift from one stable or equilibrium state of organization or developmental homeostasis to another, where stable states are equilibria points on an epigenetic landscape. Other formal models built on the same principle might represent the relationship between gene products and signaling molecules in the tissue microenvironment as vast signaling networks, reorganized by cancer. These modeling strategies are affiliated (very roughly) with some overlapping commitments or theories of carcinogenesis, broadly understood. None, however, relies exclusively on one "theory" to the exclusion of others. Rather, each draws upon various presuppositions about what sorts of causes are significant in cancer initiation and progression: mutation, epigenetics, or organization or dynamics of developmental pathways. That said, all such models more or less accept some fundamental assumptions: that cancer cells are phenotypically different from normal cells, in part due to their genetic or epigenetic changes, and in part due to many other factors at work in the tissue microenvironment. Indeed, we can identify (and in many cases have identified) the "driver" mutations and epigenetic factors responsible for these particular phenotypes.

The so-called oncogene paradigm, according to which mutations to specific genes play essential roles in the generation of a cancer phenotype at the cellular level, has dominated at least the last 30 years of cancer research. More peripheral research programs have focused attention on the role of the tissue microenvironment (Sonnenschein and Soto 1999), cellular metabolism (Warburg 1926; Seyfried and Shelton 2010), the role of structural or developmental organizing factors (Bissell et al. 1999), or stem cells (Clarke et al. 2006) in cancer initiation and progression. These focuses on different kinds of causes or causal pathways may prima facie appear inconsistent. It is my view, however, that these are not incommensurable. Instead, they are research programs focused on simply different causal pathways, all of which are indeed relevant to cancer, and they can be integrated into a more comprehensive view of cancer's origins. Models of carcinogenesis that focus on one particular local causal pathway are not fundamentally in tension with models of broader networks of pathways. For, as a matter of fact, cancer is a complex, dynamic process, requiring attention to multiple temporal and spatial scales, from short-term molecular interactions, to mid-term developmental processes shaping tissue organization, to the history of life on earth and the emergence of multicellularity.

To situate my views, then, namely the views discussed above, I do not take theoretical unification as the exclusive goal of scientific inquiry. Sometimes perspectival models are complementary; other times they are simply taken to be competitors for heuristic or exploratory purposes (Massimi 2018b). Indeed, the hope for a unified theory of cancer, if by this one understands a set of necessary and sufficient causal conditions on all cancers or universal laws of carcinogenesis, is simply misguided. Rather, there are many useful perspectives on cancer, or research programs that focus on one type of cause or on one temporal or spatial scale. Models that may appear inconsistent can be reconciled once one situates them in a larger context or interprets them appropriately. There are two components of my argument for this claim: one historical and one philosophical. These will correspond, roughly, with two parts of the paper.

First the historical: in her recent book *Cancer Stem Cells: Philosophy and Theory*, Laplane argues that cancer stem cell theory is a "revolutionary" new theory of cancer initiation and progression that offers to "break the stalemate in the war on cancer" (Laplane 2016, 2). She contrasts cancer stem cell (CSC) theory with what she calls the "classical conception" of cancer initiation and progression. She claims that the classical approach cannot explain and did not predict many important features of cancer. In particular, the classical view cannot explain and did not predict the genetic and phenotypic heterogeneity of cancer cells in a tumor, nor the causes of resilience to chemotherapy. This book echoes similar books claiming "revolutionary" approaches to cancer. In 1999, Soto and Sonnenschein argued in *The Society of Cells: Cancer and Control of Cell Proliferation* for a "new paradigm" of cancer research: the "tissue organization field theory" (TOFT), which they contrast with the somatic mutation theory (SMT). Sonnenschein and Soto contest what they take to be the universally held "dogma" that somatic mutation is the cause of cellular proliferation. In their view, cancer results from a breakdown of tissue organization that disrupts the normal inhibitions of proliferation that are inherent in the tissue architecture of a multicellular society of cells. In each of these cases a new paradigm is contrasted with the old, and promises are made on behalf of the new for both understanding and treatment of cancer. Are these alternatives genuine "paradigm" shifts, involving incommensurable views about the causes of cancer? Is the choice between purportedly competing views so stark? Some (Malaterre 2007; Bertolaso 2011) have argued that this is a mistaken way of thinking about the SMT versus TOFT debate. In this chapter, I make a similar argument in regard to a more recent debate in the history of cancer research concerning cancer stem cells. Like Malaterre and Bertolaso argue in the context of SMT versus TOFT, I argue that this debate is better characterized as a gradual shift in understanding and assimilation of novel ideas. The stem cell theory and "classical" approach are not so starkly at odds, representing incompatible theories. Indeed, in my view, progress in cancer research is not well framed as shifts in theory; cancer research is largely problem-driven as opposed to theory-driven. There are few if any paradigm shifts in science, and cancer science is no different in this respect from other cases. This is piecemeal theory change rather than replacement, and perspectival realism can shed light on how. The contrasts offered up between these different perspectives on cancer are not between incommensurable worldviews. This is my historical thesis.

Second, my philosophical thesis is intended to dovetail with the historical thesis. Searching for a unified theory or one necessary condition on carcinogenesis is exactly the wrong strategy. Contrary to what defenders of these "revolutionary" new theories suppose, cancer is not either a disease of mutations *or* a disease of the tissue microenvironment, a disease of genes *or* a disease of stem cells. Rather, each of these research programs provides a novel but *partial perspective* on a complex, heterogeneous disease. Each approach has shed light on the mechanisms that yield cancer, though emphasizing quite different temporal and spatial scales.

The view I defend here, in other words, may be characterized as a kind of theoretical pluralism. According to Beatty, this is the view that a domain of inquiry is

> essentially heterogeneous, in the sense that a plurality of theories or mechanisms is required to account for it. . . . There is no single theory or mechanism—not even a single, synthetic, multi-causal theory or mechanism—that will account for every item in the domain.
>
> (Beatty 1995, 65)

Beatty has suggested that we ought to expect theoretical pluralism in the biological sciences. His rationale is as follows:

> Why should we adhere to a methodology that dictates the search for unitary accounts of each domain of biological phenomena—e.g., a unitary account of inheritance, or a unitary account of carbohydrate metabolism, or a unitary account of gene regulation, or a unitary account of speciation—unless we have reason to believe that the outcomes of evolution are *highly* constrained? . . .
>
> . . . Unless we believe the outcomes of evolution are always severely constrained, then perhaps we should be on the lookout for multiple accounts in each domain.
>
> (Beatty 1995, 75)

Beatty thinks that unless we have special reason to think that a biological process is evolutionarily constrained, we ought to seek out not unitary but pluralistic accounts of phenomena. Is the pathway to cancer constrained either from an evolutionary perspective or otherwise? In some sense the pathway is constrained, but this is only insofar as all cells in a multicellular organism are the product of a long history of evolution of cooperative organization. However, constraints on disruptive growth have evolved in different ways in different tissues (and, indeed, in different sexes and different species!); many different constraints on cancer have been selected for, and there are also many ways in which these constraints fail. This is why cancer is, after all, not one disease; each cancer has its own distinctive site of origin and so also its own distinctive pattern of failure, unique genetic signature, pattern of progression, and likely outcomes—as well as, of course, its distinct remote and proximate causes, from viral infection to environmental factors. Cancers are heterogeneous in a variety of senses, both genetic and phenotypic, or if you like, distinct in ontogeny and phylogeny. At only the most coarse-grained level of description is there one way in which a cell becomes a cancer cell, a cancer cell becomes a population of cancer cells, and a population of cells invades and metastasizes to neighboring tissues.

The right lesson to take away from the history of cancer research, in other words, is that the question of which among many possible research programs is the most "unified" or gives the "true theory" is simply the wrong question to ask. Giere's perspectival realism is useful here: in the face of complexity, making progress in science is not a matter of searching for one true theory or view from nowhere. The right way to consider the problem is to note how and why different research traditions are epistemically fruitful—where this means that they yield knowledge of causes or properties of the system that help us better understand, predict, and successfully intervene.[6] Now I will turn to my historical analysis.

2 SMT versus TOFT

In 1999 Soto and Sonnenschein published *The Society of Cells*, in which they set out what they take to be incommensurable views of carcinogenesis. On what they take to be the widely accepted and yet false view—the somatic mutation theory (SMT)—mutations are acquired during somatic cell division by the precursors of cancer cells. Some such mutations yield the hallmarks of cancer—cancer cells proliferate, are not sensitive to apoptotic signaling (signals that indicate cells should become senescent), acquire a blood supply, and acquire the capacity to invade and metastasize. This picture of cancer, they argue, presupposes that the default state of metazoan cells is quiescence—cancer is a departure from this default, and the explanation for the departure from this state is, in their view, "reductionist." In contrast, they advance what they take to be a holist view, the "tissue organizational field theory" (TOFT). According to TOFT, the default state of cells is proliferation, and the cause of cancer is disruption of reciprocal interactions between cells that ordinarily serve to maintain tissue organization. They argue that their picture of cancer can explain phenomena that SMT cannot; in particular, it explains and predicts the fact that cancers exhibit a great deal of heterogeneity, and that it is possible, by altering the tissue microenvironment, for cancer cells to revert to healthy, normal cells and develop into differentiated tissue. Therefore, in their words, we ought to adopt this novel paradigm and reject the false and failed alternative.

There are two components of my argument; one historical and another conceptual. First, it is not clear to me that any advocates of the somatic mutation theory (whom Soto and Sonnenschein only rarely identify) ever were committed to the view they describe about the default state of cells, or for that matter, that a single or few oncogene(s) may induce cancer. In other words, it seems to me that Soto and Sonnenschein give at best a very thin caricature of mainstream cancer researchers' views. Second (perhaps not surprisingly), once we get a more robust picture of the mainstream view, we need not take these views as incompatible. Indeed, many of the classic studies that Soto and Sonnenschein cite in support of their novel theory are not only consistent with the presence of mutations in cancer cells but also mutually reinforcing. There are reciprocal interactions between tissue organization and signaling pathways controlled in part by many mutations to cancer cells. By couching the issue as revolutionary, and so enforcing a sharp divide between precursors and successor paradigms, one is bound to view history as a vanquishing rather than a progressive adding of novel perspectives on the same phenomena.

3 CSC Theory: Cancer Stem Cell Theory

In her recent book, the philosopher Laplane argues that the CSC theory has various advantages over the alternative classical theory (Laplane

2016). Tellingly, she does not attribute the classical theory to any particular author or set of authors. In fact, she defines the classical theory primarily in terms of how it differs from CSC theory. So it may be helpful first to describe what she takes to be the four fundamental theses of CSC theory:

1. CSCs are capable of self-renewal, thus producing new CSCs.
2. CSCs are capable of differentiation, thus producing cells of different phenotypes.
3. CSCs represent a tiny subpopulation of cells, distinct from other cancer cell populations, and are in theory isolatable.
4. CSCs initiate cancers.

It is worth noting that the first two claims, as she points out, concern the concept of a stem cell. The latter two claims concern carcinogenesis itself: how cancer arises. It is the latter two theses that Laplane takes to be in tension with the classical view. On the classical theory, she claims that "all cells are capable of self-renewal" (Laplane 2016, 33) and "different cell types are able to initiate new tumors" (Laplane 2016, 33). The capacity to initiate new tumors arises as a consequence of the acquisition of random mutations, what she calls the "stochastic model." This same capacity—the acquisition of random mutations and the evolution of cell lineages—also is taken to explain the heterogeneity of cell populations and the capacity for cells to acquire resistance to chemotherapy. She takes these additional commitments of the classical theory to be ad hoc and contrasts this failure in "parsimony" of the classical view with her preferred CSC theory.

That is, she claims that CSC theory has the "ability to explain various phenomena (cancer development and propagation, as well as relapse) from a very limited number of hypotheses" (Laplane 2016, 28). In contrast, the classical view neither predicts nor explains these phenomena but must invoke special ("additional" or "ad hoc") theories to explain them. The CSC is thus more parsimonious than classical theory because it unifies a number of explanations, or shows how different phenomena can be explained by a common unified theory. In particular, the low clonicity of cancer cells and high heterogeneity of tumors is best explained by CSC. In contrast, the classical theory must invoke many different additional hypotheses. Thus the CSC theory is more parsimonious than the alternative classical theory. Moreover, the CSC theory has the advantage that it connects "basic research to biomedical interventions by suggesting a new therapeutic strategy for cancers" (Laplane 2016, 28).

It is not the case that these two theories are as a matter of fact inconsistent once we begin to explore a point that Laplane herself draws our attention to: namely, that the concept of cancer stem cell is multiply ambiguous. Indeed, there are different variants on the same general theory that disagree on one specific point, namely, the origins of cancer stem cells. To explain:

if, as a matter of fact, all cells at least potentially may acquire the properties typical of cancer stem cells (which is an independent empirical question, the evidence for which is still being gathered), then the CSC theory is perfectly consistent with the classical theory. Indeed, some populations of cells in a tumor appear to all have the features of a cancer stem cell or the potential to behave like a stem cell. That is, "stemness" is a property associated with certain capacities that are not fixed but acquired. The plasticity of many types of cancer cells makes it the case that many cancer cells can shift back and forth between "stemness" phenotype and non-stem phenotype. Given this, it appears that the CSC is just one of a continuum of general views, some of which take only specific types of cells to be precursors to cancer and others that grant that many different types of cells have the potential to develop such properties. But this points to a more substantial issue, one she herself is at pains to defend: stemness itself is a relatively unstable category in the cancer literature. Is stemness just a proxy for whatever properties there are that allow a cancer to arise? Is having such properties just what it *means* to be a CSC? For if a CSC is just any cell that initiates a tumor, then CSCs *must* exist (something must be initiating a tumor!) and the classical theory *must* endorse the existence of CSCs. It is by definition true that cancer stem cells exist if cancer stem cells are just those cells that initiate tumors. So the real question at issue here is whether the cells that initiate a tumor are in some way distinctive or require distinctive precursors. But classical theories of carcinogenesis of course granted that the cells that initiate a tumor must possess a variety of features that make them distinctive. The real question is what features those are, which (at least initially) was an open question on the classical theory. So it is unclear, then, what the purported disagreement is about.

Moreover, the CSC theory is not as parsimonious as Laplane makes out initially in chapter 2. As she later acknowledges in chapter 5, the CSC theory can (and indeed must) help itself to the somatic evolution theory if it is to explain a variety of features of cancer development and metastasis. So this extra, or additional, hypothesis that renders the classical theory less parsimonious is one that the defender of the CSC theory (eventually) endorses as well.

The real innovation of cancer stem cell theory, in my view, is in giving a label to something that classical theory already acknowledged as a legitimate and even likely possibility. Namely, there are special or unique features belonging to all and only those cells that initiate a tumor and/or cells that propagate tumors or yield metastases. It is, after all, still an open empirical question whether any cell in the body is capable of acquiring these properties or only some. In other words, what is at issue between the two is whether all cells have the potential to acquire the properties of those cells that can initiate and propagate cancers. But this is a matter of debate within the CSC literature. So it's perfectly possible for defenders of the classical view to endorse (at least one version) of the CSC. The two are not so starkly opposed as Laplane makes out. As Laplane documents at some length, the very concept of a cancer stem cell is multiply ambiguous, in the following ways:

- First, when we speak of cancer stem cells, we may be referring to their capacities or properties, or to their historical role or genealogy, that is, to the fact that they were the cells from which other cancer cells originate. That is, some take cancer stem cells to be defined in terms of their distinctive capacities and some in terms of their relationship to other cells—in particular, to their ancestor-descendent relationships in a population of cells in a tumor. The "cancer stem cell model" is sometimes simply taken to refer to any model of a tumor that treats the population of cells as having a hierarchical relationship, where one or a few cells propagate the tumor, whether or not those cells have distinctive properties that cause them to stand in that relationship.
- Second, there are several different kinds of historical role that CSCs might play: they may be all and only those cells that initiate a cancer under natural conditions, they may be those cells which propagate a cancer in situ, or they may be those cells that are capable of propagating a cancer in an experimental animal.
- Third, some take the concept of CSC to be restricted to normal stem cells, which some believe are the most likely precursors to cancer. Others hold that cells that originate a tumor have stem-like properties but may or may not derive from normal stem cells.

Cancer researchers have attempted to give greater clarity to the debates about cancer stem cells by using different terminology to distinguish between these different senses: "cancer-initiating cells," "cancer-propagating cells," "cancer stem-like cells," and so on. But in Laplane's view, none of these attempts at clarification did the work the authors hoped. For it turns out that even the expression "cancer-initiating cells" could refer to either precancerous cells that have acquired some but not all of the properties necessary to initiate a tumor, cancerous cells that initiate tumors in patients, or cancerous cells that initiate tumors in experimental animals. All three senses have been used in the literature, leading to some confusion.

Here's where it becomes clear that Laplane has set up a false dichotomy. She points out that different experimental conditions can lead to different results in the propagation of cancers in experimental animals. Under some conditions, much higher percentages of cells in a tumor are capable of propagating a cancer in experimental animals. Whereas the initial experiments in propagation yielded a very small success rate—only 0.00001 percent of cells in leukemia—as many as 25 percent of cancer cells from a melanoma could propagate themselves in NSG mice (mice where a gene associated with the precursors to natural killer cells was disabled). Indeed, using different mice strains, and even different sexes of mice, yields greater or lesser success at propagation by a much higher percentage of "CSC" cells. Instead of viewing this as evidence in favor of the classical theory—namely, that any number of cells is capable of acquiring the features necessary to propagate a tumor—she suggests only that this evidence undermines "the idea that CSCs only represent a small fraction of cancer cells" (Laplane 2016, 94). But if CSCs are just any cell capable of propagation, this is by definition true. The real question at issue is whether any cell can acquire this capacity in the right circumstances or whether only some can. And this question is not definitively decided by such experiments, although they do lend greater credibility to the classical model than Laplane acknowledges. In other words, it sometimes seems that Laplane, despite the fact that she acknowledges that the very expression "CSC" is multiply ambiguous, fails to recognize that it is this very ambiguity that leaves the door open to seeing the CSC theory and classical theory as overlapping and fully consistent perspectives.

4 Conclusion

How may this account of the recent history of cancer research be brought to bear on the debate over perspectival realism? Recall the taxonomy of relationships between models discussed by Fagan (2017):

- Direct conflict
- Simple additivity
- Subsumption
- Interactive process
- Cross-perspective translation
- No cumulative interaction
- Non-interaction
- Complementarity.

How can this picture help us make sense of debates among advocates of purportedly competing theories about cancer? We can see these alternatives as several ways in which competing "perspectives" can be reconciled. Of course, different authors mean different things by "perspectives." For Giere, a perspective is akin to Kuhn's disciplinary matrix; for Massimi, it is the scientific practice of a given community; for Teller (2001), it is a family of idealized and imprecise models. In this context I take a perspective to be a family of commitments regarding what causes are central or important to cancer, associated with a heuristic or framework, which helps us develop research questions, frame appropriate answers, and guide inquiry into cancer. All of these together make up a perspective. Different research programs have focused attention on different temporal and spatial scales or concerned themselves with one or another causal pathway as central to cancer initiation and progression. These in turn provide us with principles to build models of cancer initiation and progression. The oncogene paradigm led to the identification of a variety of genes, the mutation of which led to uncontrolled growth, failure of apoptosis, angiogenesis, and eventually invasion and metastasis. Early models of cancer growing out of this research tradition represent cancer progression as a stepwise, rate-limited acquisition of a series of mutations, eventually leading to uncontrolled growth.

Models growing out of "competing" perspectives or research traditions focused on the roles of tissue microenvironment and tissue architecture in cancer, the typical features and behaviors of stem cells in cancer, or the developmental pathways disrupted by or co-opted in cancer progression. For instance, one model that draws upon the cancer stem cell theory takes it to be the case that the differential incidence of cancers of different tissue types is largely due to the number and rate of division of somatic stem cells in different tissue types, given the relatively strong correlation between the two (Tomasetti and Vogelstein 2015). But this model is not in tension with the classical model. In fact, both models treat cancer as a stepwise, iterated, and rate-limited process, where mutations and epigenetic alterations eventuate in disease.

In some cases, these different models of cancer are concerned with different outcomes or classes of outcome at different scales. They focus on different causal pathways to cancer or are concerned with different scales of analysis (from the molecular on up to evolutionary history). So on the one hand, we might say that these models are complementary and non-interactive; they are not in conflict, insofar as they are concerned with different questions. However, in other cases, several models have been developed for describing progression to disease, the dynamics of progression, or subsequent metastasis within a single cancer type or subtype, such as breast cancer. In these cases, there appears to be a relatively seamless integration of theory and data with mutual constraint, drawing upon evolutionary and developmental perspectives, knowledge about metabolic changes to cancer cells, and structural and developmental factors in cancer, genetics, and stem cell theory. Indeed, arguably, the classic multistage model predicts that the hierarchical structure of differentiation in tissue is a protective mechanism against cancer and thus serves as a kind of anticipation of stem cell theory. For if a normal self-renewing population of stem cells acquires mutations or epigenetic changes that yield increases in proliferation or resistance to apoptosis, then they can yield a cancer via somatic evolution (Pepper, Sprouffske, and Maley 2007). According to the current stem cell theory, stemness properties could either be a defining feature of some subpopulations of cells in a tumor or could be a transitory property of all cells in a tumor. For cancer cells appear to be highly plastic and can transition back and forth between stem and non-stem states (Kreso and Dick 2014). Ultimately, however, genetic changes, epigenetics, and changes to the tumor microenvironment all contribute to the emergence of disease. These perspectives are not in tension but complementary; and seeing how and why they are mutually informative has been a progressive, gradual process. The integration of theory and data is iterative, as more information about the various properties that contribute to cancer progression, heterogeneity, and resistance to chemotherapy, and their mechanistic bases, is acquired (Plutynski 2013). Massimi's interactionist approach seems the best fit here. Different models of the same cancer or cancer subtype (e.g., breast cancer) that focus on different causal pathways each relevant to the larger outcome can be seen as yielding complementary information about constraints on this process.

The attempt to tell this story as one of vanquishing the old and replacing with the new is, in my view, a mistake. This model of successive theory vanquishing, or of the replacement of one incommensurable paradigm with another, is inappropriate here and leads to unproductive battles. Instead, what has occurred is a progressive integration of diverse perspectives on the same phenomena or alternatively, in some contexts, the development of models concerned with slightly different, and equally important, questions or problems, or different targets of inquiry.

Notes

1. Or at least this is one very influential reading of Kuhn, though one Kuhn protested (1977).
2. There are, as a matter of fact, several "middle-ground" perspectives that have been offered, some to the effect that the debate proposes a false dichotomy. See, e.g., Stein (1989); Fine (1984).
3. Giere's proposal was initially intended as a middle ground between extreme versions of "objectivist" scientific realism (the thesis that theories can in principle provide "a complete and literally correct picture of the world itself"; Giere 2006b, 6) and constructivist antirealism ("scientific claims about any reality beyond that of ordinary experience are merely social conventions"; Giere 2006a, 26).
4. There are, of course, a variety of competing views about what it means to say that the world is in some sense "complex." For a discussion, see, e.g., Wimsatt (1994) and Mitchell (2003). Perhaps it is needless to say that not all agree that the world (in general) is complex or in what sense(s). My own view is that perhaps only some types of systems exhibit both what Simon (1962) calls "organizational" and "dynamic" complexity (e.g., organisms, beehives, cities).
5. I take it that some families of models are informed by a research tradition that includes a commitment to certain claims as well supported by evidence, but nothing like a set of laws or "theoretical" principles. It is unhelpful to speak of "theories" in this context, at least in the sense of lawlike, exceptionless generalizations about cancer initiation or progression of the sort that philosophers (at least historically) have identified with laws of nature. Instead, going back as far as Virchow's (1863) proposal that cancer may result from irritation, it is rarely (if ever) the case that cancer scientists assume or confidently

assert (unless they're being incautious or writing for a popular audience) that any particular distal cause or proximate mechanism is a necessary condition on cancer. Instead, over the course of the history of cancer research, viruses, "oncogenes," "tumor suppressor" genes, metabolic changes to cells, or "stemness" properties are taken to be highly probable, plausible, or likely candidate causes of cancer initiation, progression, or recurrence.

6. Morange (2015) makes a similar (but more general) point. He argues that

> in some cases . . . contrasts (between competing perspectives) are hardened by participants. Both sides demand that a choice be made between the different explanations. In other cases, the need for a choice vanishes when knowledge of the system under study increases . . . the discontinuity (between approaches) is progressively disappearing.
>
> (Morange 2015, 40–41)

The latter is very much the case in cancer research, in my view.

References

Beatty, J. 1995. "The Evolutionary Contingency Thesis." In *Concepts, Theories, and Rationality in the Biological Sciences*, edited by Wolters, G., and Lennox, J., 45–81. Pittsburgh: University of Pittsburgh Press.

Bertolaso, M. 2011. "Hierarchies and Causal Relationships in Interpretative Models of the Neoplastic Process." *History and Philosophy of the Life Sciences* 33(4): 515–535.

Bissell, M. J., Weaver, V. M., Lelièvre, S. A., Wang, F., Petersen, O. W., and Schmeichel, K. L. 1999. "Tissue Structure, Nuclear Organization, and Gene Expression in Normal and Malignant Breast." *Cancer Research* 59(7 Supplement): 1757s–1764s.

Chakravartty, A. 2010. "Perspectivism, Inconsistent Models, and Contrastive Explanation." *Studies in History and Philosophy of Science* 41(4): 405–412.

Chirimuuta, M. 2016. "Vision, Perspectivism, and Haptic Realism." *Philosophy of Science* 83(5): 746–756.

Clarke, M. F., Dick, J. E., Dirks, P. B., Eaves, C. J., Jamieson, C.H.M., Jones, D. L., Visvader, J., Weissman, I. L., and Wahl, G. M. 2006. "Cancer Stem Cells—Perspectives on Current Status and Future Directions: AACR Workshop on Cancer Stem Cells." *Cancer Research* 66(19): 9339–9344.

Fagan, M. B. 2017. "Explanation, Multiple Perspectives, and Understanding." *Balkan Journal of Philosophy* 1: 19–34.

Fine, A. 1984. "The Natural Ontological Attitude." In *Scientific Realism*, edited by Leplin, J., 261–277. Berkeley: University of California Press.

Giere, R. N. 1999. *Science Without Laws*. Chicago: University of Chicago Press.

Giere, R. N. 2006a. *Scientific Perspectivism*. Chicago: University of Chicago Press.

Giere, R. 2006b. "Perspectival Pluralism". In *Scientific Pluralism, Minnesota Studies in the Philosophy of Science XIX*, edited by Kellert, S. H., Longino, H. E., and Waters, C. K., 26–41. Minneapolis: University of Minnesota Press.

Kreso, A., and Dick, J. E. 2014. "Evolution of the Cancer Stem Cell Model." *Cell Stem Cell* 14(3): 275–291.

Kuhn, T. S. 1977. *The Essential Tension*. Chicago: University of Chicago Press.

Kuhn, T. S. 2012. *The Structure of Scientific Revolutions*, 4th ed. Chicago: University of Chicago Press.

Laplane, L. 2016. *Cancer Stem Cells*. Cambridge, MA: Harvard University Press.

Malaterre, C. 2007. "Organicism and Reductionism in Cancer Research: Towards a Systemic Approach." *International Studies in the Philosophy of Science* 21(1): 57–73.

Massimi, M. 2018a. "Four Kinds of Perspectival Truth." *Philosophy and Phenomenological Research* 96(2): 342–359.

Massimi, M. 2018b. "Perspectival Modeling." *Philosophy of Science* 85(3): 335–359.

Maxwell, J. C. 1881. *A Treatise on Electricity and Magnetism. Volume 1*. Oxford: Clarendon Press.

Mitchell, S. D. 2003. *Biological Complexity and Integrative Pluralism*. Cambridge: Cambridge University Press.

Morange, M. 2015. "Is There an Explanation for . . . the Diversity of Explanations in Biological Studies?" In *Explanation in Biology*, edited by Braillard, P., and Malaterre, C., 31–46. Dordrecht: Springer Netherlands.

Morrison, M. 2011. "One Phenomenon, Many Models: Inconsistency and Complementarity." *Studies in History and Philosophy of Science* 42(2): 342–351.

Pepper, J. W., Sprouffske, K., and Maley, C. C. 2007. "Animal Cell Differentiation Patterns Suppress Somatic Evolution." *PLoS Computational Biology* 3(12): e250.

Plutynski, A. 2013. "Cancer and the Goals of Integration." *Studies in History and Philosophy of Science Part C: Studies in History and Philosophy of Biological and Biomedical Sciences* 44(4): 466–476.

Seyfried, T. N., and Shelton, L. M. 2010. "Cancer as a Metabolic Disease." *Nutrition & Metabolism* 7(1): 7.

Sonnenschein, C., and Soto, A. M. 1999. *The Society of Cells*. Oxford: Bios Scientific.

Stein, H. 1989. "Yes, But . . . Some Skeptical Remarks on Realism and Anti-Realism." *Dialectica* 43(1/2): 47–65.

Teller, P. 2001. "Twilight of the Perfect Model Model." *Erkenntnis* 55(3): 393–415.

Tomasetti, C., and Vogelstein, B. 2015. "Variation in Cancer Risk Among Tissues Can Be Explained by the Number of Stem Cell Divisions." *Science* 347(6217): 78–81.

Virchow, R. 1863. *Die krankhaften Geschwülste*. Berlin: Hirschwald.

Warburg, O. H. 1926. *Über den Stoffwechsel der Tumoren*. Berlin: Springer.

Wimsatt, W. C. 1994. "The Ontology of Complex Systems: Levels of Organization, Perspectives, and Causal Thickets." *Canadian Journal of Philosophy* 24(Supplement 1): 207–274.

10 Perspectives, Representation, and Integration

Sandra D. Mitchell

1 Introduction

Contemporary science studies complex structures and behaviors at a variety of levels of organization, from the most basic component parts to the entire system, using representations of different degrees of precision, from fine- to coarse-grained. The relationships among these multiple scientific models continue to be in dispute. Reductionists argue that all explanations and predictions in principle, if not in practice, can be crafted from descriptions of the properties and behaviors at the most fundamental level of physical substance. The argument presumes that since all more complex arrangements are composed from the basic physical building blocks, knowledge of the basic level could or will eventually tell us everything about composite structures and their behaviors. Thus, if this were the case, all causal information that appears in higher-level models should be recoverable in models of the lower or lowest level. Yet some coarse-grained explanations or predictions that appeal to higher-level structures seem to be required to explain at least some features of complex behaviors.[1] I have argued (Mitchell 2009) that this reductionist argument confuses compositional physicalism with "physics-ism." The first, compositional physicalism, is a metaphysical presumption that is constitutive of scientific practice, while the second, physics-ism, is a claim about scientific representations, one which is unsustainable. Scientific representations do not mirror the compositional relations of natural systems; rather, they encode perspectives.

This chapter explores some consequences of the perspectival nature of representation. Following Giere's (2006a) seminal discussions, I argue that if representational models are both partial and perspectival, then in order to acquire knowledge of natural systems, science must employ a plurality of models, methods, and representations. If perspectival pluralism is correct, a further question arises. What are the relationships among multiple models and representations of the "same" phenomenon? I will argue that perspectivism gives us resources for understanding those relationships in new ways. I will provide a detailed example of contemporary, experimental models in structural biology that both exemplifies the partial, perspectival character of models in general and the epistemic value of pluralism. In section 2, I will present the argument from perspectivism to pluralism. In section 3, I illustrate how perspectivism works in the case of modeling protein folding. In section 4, I will consider the possible relations among multiple models and philosophical resources for understanding those relations. I will then argue for the epistemic value of integrative strategies for managing a plurality of models of a single phenomenon.

2 Perspectivism and Pluralism

Giere has identified several important features of scientific perspectivism, a view he casts as an alternative to both objectivist and constructivist accounts of scientific models. In his words:

> general principles by themselves make no claims about the world, but more specific *models* constructed in accordance with the principles can be used to make claims about specific aspects of the world. And these claims can be tested against various instrumental perspectives. Nevertheless, all theoretical claims remain perspectival in that they apply only to aspects of the world and then, in part *because* they apply only to some aspects of the world, never with complete precision. The result [is] an account of science that brings observation and theory, perception and conception, closer together than they have seemed in objective accounts.
>
> (Giere 2006a, 15)

For Giere, theories and their associated models represent only some aspects of the phenomenon studied and do so with something less than perfect precision.[2] This is in contrast to a "mirror of nature" completeness in which every feature of a phenomenon is precisely described in a one-to-one mapping into a model. Indeed, no representational model can attain this type of completeness without it being an exact, to-scale copy of the target phenomenon. Cluttering the lab with duplicates of the phenomenon studied would fail to bring science much closer to explaining and predicting natural behavior than directly engaging with the original. Complete models, in this sense of completeness, would be of no scientific use. To be useable, a representational model has to leave some things out by abstraction or simplify by idealization. That is, scientific models are partial and imprecise.

The flip side of partiality is that by leaving some features out, every method, model, and representation "selects" features to be included. This "selection" reflects a perspective. Is this the result of deliberative choice?

Many have argued that models and representations have an intentional component. Giere summarizes this as "agents 1) intend; 2) to use model, M; 3) to represent a part of the world, W; 4) for some purpose, P" (Giere 2010, 269). Suárez (2010) presents a comprehensive overview of the different positions regarding the role of intentionality in both the analysis of what constitutes representations and in how representations are used in scientific practice.

I agree with both Giere and Suárez (see also van Fraassen 2008) that for something to *be* a model representing some aspect of nature, an agent must use it for that purpose, otherwise it is just cardboard and glue, ink on paper, or pixels on a LCD screen. But which features are left out and which are kept in may not always be explicitly intended. What needs further explication, and what contributes to the content of a perspective, is how and what particular features are selected to be represented in the model. Part of the answer may come from the intentions of those developing models with a goal in mind, but part may be a function of the methods and medium themselves. What features at what degree of precision can an instrument detect? The experimental model that is inferred will reflect the methodological constraints. What form of representation of the model is used to communicate its predictive and explanatory content? Graphical, pictorial, and mathematical forms, among others, also impose constraints on what can be depicted (Perini 2012). For example, as I will detail in section 3, the variety of scientific models developed to predict the structure of a functional protein vary in the key features represented, in the methods for generating a prediction, and in the idealized contexts in which the models most directly apply. But before doing so, let us take a closer look at the relation between perspectivism and pluralism about models.

Partiality and perspectivism entail model pluralism. A single model cannot deliver a complete, maximally precise representation of a given aspect of nature. What it leaves out could be, and often is, represented by other perspectival models. If the features that are left out in one model but included in another are causally independent, partitionable into distinct subfeatures, or neatly mereologically nested, then the multiple models might be simply combined to form a single, more complete model of the phenomenon. If they are not, then a plurality of models is entailed by the partiality of representation. Since a single model cannot deliver all the causally relevant aspects of a given phenomenon with complete precision, using multiple models may be required to be adequate to the explanatory or predictive goal. I will show below that integrating multiple, compatible models can increase scientific knowledge of nature. While the enduring plurality of models cannot be reduced or unified to produce a single model, they can be interactively integrated, yielding increased accuracy while retaining perspectival pluralism. This is the situation I will explore in the rest of the chapter.

Different scientific perspectives are characterized by different assumptions, methods, instruments of observation, experimental arrangements, concepts, categories, and representations, all of which are associated with specific pragmatic concerns and explanatory or predictive projects. How a natural phenomenon "looks" to one perspective is different from the way it "looks" to another. Consider by analogy how the world looks to beings that have different visual systems—a bee and a human, for example. Bees and humans are trichromatic; that is, they each have three photoreceptors within the visual system by which color perception is constructed. Humans base color combinations on red, blue, and green wavelengths, while bees base all their colors on ultraviolet, blue, and green. Thus the same flower from the perspective of a bee and the perspective of a human looks very different (Giere 2006b; Chirimuuta 2016). It is the same flower made of the same material in the same environment, but the visual apparatus of a bee and the visual apparatus of a human access different visual signals that are afforded by the same flower.

Or consider how sensory modalities are integrated in human beings, as another example. Our five senses—sight, sound, taste, hearing and touch—permit us to acquire information about nature. These different modalities detect different aspects of a given phenomenon. Again, perspectivism and partiality are evident, but here they are not choices we make—rather they are what types of signals are accessible by a given sensory apparatus. Consider sight and sound. What information can we acquire visually? Reflectance, saturation, color, and light reflection and refraction. Simply put, the human eye has a cornea—like a camera lens—that focuses light onto the retina. The retina includes millions of light sensitive cells, rods, and cones. When light hits the rods and cones, it is converted into an electrical signal that is relayed to the brain's visual cortex via the optic nerve. What about sound? What auditory information do we acquire? Now it is sound waves, not light, that is detectable. The basilar membrane in the inner ear detects frequencies of sound waves by vibration. Different frequencies activate different groups of neurons on this membrane. In addition to detecting what tone is being emitted by the target source, hearing also can locate the source of the sound by using the difference in loudness and timing between the two ears. As the eye and ear illustrate, the same organism can have multiple apparatuses to detect different features or different aspects of the same feature of what might be spatiotemporally identified as the same phenomenon. The auditory and visual models of the relative location of an entity, for example, may be different reflecting as a result of the causal interaction of signals from the source with the detecting apparatus.

For most tasks in which we engage something in the external world, we employ multiple senses. Humans use vision to see an object, detecting its color, shape, and so forth relative to background and foreground to infer from visual cues the distance of the object from the observer. The observer can also use hearing to discriminate sounds coming from the object from background noise. Tactile cues and olfactory cues can also be detected by touch and smell. The representational resources vary with the sensory modality. An object affords different, but equally accurate, representations through these different sensory modalities. The individual senses may have optimal usefulness in different circumstances. Not surprisingly, experimental results have shown that the collective use of multiple modalities, like vision and audition, increases the likelihood of detecting and identifying events or objects above those of only one. What might be surprising is that the degree to which multisensory integration is better is superadditive. Multimodal integration in the brain has been studied at both the neuronal level and by the results of experiments on task completion: "the integrated product reveals more about the nature of the external event and does so faster and better than would be predicted from the sum of its individual contributors" (Stein and Stanford 2008, 255). So not only does using different perspectival senses provide more information than using a single modality, the cross-modal interaction of stimuli can lead to multisensory integration, which yields a nonlinear, superadditive neuronal response and faster than computationally additive time to task completion.

I suggest that these aspects of sensory perspectivism are analogous to how different scientific models represent the same phenomenon. They use different theoretical or experimental apparatus that access different aspects of the phenomenon. The conceptual framework, particular preconceptions as well as the representational medium, and the problems pursued and the methods for answering them all vary between different scientific perspectives. Multiple models from different perspectives can be used together, in non-unifying and non-reductive ways, to explain or predict the same phenomenon. A constitutive assumption of scientific modeling is that there are phenomena in nature independent of us. However, all scientific engagement with phenomena is refracted through the lenses of different modes of causal interaction (via experience and experiment, akin to vision or audition) and conceptual, mathematical, and propositional representations.

Take the example of a protein as a phenomenon in the world. A protein is classified by its linear polypeptide chain of amino acids (produced in sequence in the cell from the coding information of DNA). The amino acid sequences can realize up to four levels of structure. The first, the linear amino acid sequence, can form into a secondary structure of a sheet or a helix by means of hydrogen binding. The tertiary structure describes the overall three-dimensional shape of the entire protein including secondary structures plus the linear structure. This conformation includes bends and twists in irregular patterns formed by the bonding interactions of the side chains of the amino acid components. Some of the binding sites on the protein can be buried in the interior of the structure, while others are exposed on the exterior. Thus the conformational structure that a protein takes permits or prevents binding with other molecules, thereby allowing it to perform specific biological functions. Some biological functions require larger, stable aggregate units or protein complexes composed of multiple proteins or of proteins and other molecules. This is the quaternary level of structure. Tertiary structure (or, sometimes, quaternary structure) is the functional unit of biological activities, such as oxygen transport (hemoglobin), light signaling for vision (rhodopsin), and immune response (cytokines). The protein structure scientists (and drug design engineers) aim to discover is the stable, functional structure of the protein. Scientists predict that structure from within different perspectives. So although there is some objective phenomenon in the world—the tertiary structure of a protein—it can appear differently when refracted through different scientific perspectival models.

As I will detail below, ab initio models and varying experimental models can display different locations and relations of the atomic components of the very same protein. Is at most one of these true? Are the others false? If they differ, are the models inconsistent with each other?

Defenders of model perspectivism have appealed to the non-propositional representational relationships found to illustrate some of the features of scientific representations that are not adequately captured by treating models as truth-bearing descriptive statements about the world. As Giere puts it, "strictly speaking, it makes no sense to call a model itself true or false. A model is not the kind of thing that could have a truth value" (Giere 2006a, 64). Instead, the model-world relationship is one of similarity, not a matter of truly describing all features with perfect precision (Teller 2001). Although models support propositions, they are not themselves propositional. For the same reason, it follows that model-model relationships cannot be said to be consistent or inconsistent. Logical vocabulary appropriately attaches to propositions, not to models.

Maps are non-propositional modes of representation that have been used to shed light on scientific model-world relationships. Maps encode relational, typically spatial, information that can be isomorphic, homeomorphic, or otherwise similar to the part of nature it maps. Consider multiple maps of the same place, say a street map of intersecting and parallel lines, which by convention we interpret as representing the relative location of different roads, and a topographic map, which by varying circular lines or shaded areas different elevations may be represented and interpreted. Both maps can be accurate in depicting the features they represent, but they differ in precision, scale, and adequacy to serve a specific purpose. A street map is not true or false, nor more or less realistic than a topographical map. Two street maps might be compared for accuracy: if one puts the freeway to the north of the river and the other puts it to the south, then empirical evidence will determine which better fits the location of the freeway. But purpose or goal will determine the adequacy of a map, not its accuracy. Appeal to maps reveals how multiple scientific models of the same phenomena can be equally accurate but depict different features and vary in adequacy relative to a given purpose. This is an example of how multiple partial, perspectival representations can provide greater resources for different purposes than trying to unify or reduce all the information into one, even if it is at the finest granularity of representation the medium allows. Superimposing a topographical map onto a street map, for example, would make it less useful for driving from the city center to the botanical gardens, although it might help those who want to walk or bicycle there.

While both reductive and abstractive strategies have significant roles to play in science, they are not always, perhaps not even often, sufficient on their own to capture the ways in which science works. In particular, for complex behaviors, where the phenomena display multiple components, multiple causal factors, emergent properties, robust dynamics, and more, unitary strategies are most likely to fail (Bechtel and Richardson 2010). Integrative pluralism provides an alternative picture of the plurality of models. Integrative pluralism recognizes the multiple epistemic sources of the theoretical, explanatory, and instrumental pluralism that characterizes scientific practice. However, it is not equivalent to "anything goes" pluralism (Feyerabend 1970), as empirical evidence continues to serve as a methodological foundation for the acceptance or revision of scientific beliefs. In addition, scientific models are better described in terms of compatible differences or competing alternatives than as inconsistent or incommensurable theories (Mitchell 2000, 2003, 2009).[3] For example, single idealized causal models are often developed to explain what part of a result is due to a particular cause. Genes and environment both contribute to all traits of an organism. In complex traits, like psychiatric disorders, genetic factors have been identified as explaining familial patterns, for example in major depressive disorder. Similarly, environmental pathogens, like childhood trauma from abuse or the loss of a parent, also contribute causally to the incidence of major depressive disorder. There are studies that indicate that genes moderate responses to environmental factors (Caspi and Moffitt 2006; Tabery 2014). Caspi and Moffitt showed that individuals with one or two copies of the 5-HTT short allele (a polymorphism in the promoter region of a serotonin transport gene) were more likely to suffer from depression than those with two copies of the long allele when those individuals experienced stressful events. But it is also the case that environmental factors moderate gene expression (Robinson, Grozinger, and Whitfield 2005). Caspi and Moffitt suggest that the epidemiological perspectives of gene-environment interactions are limited, and "therefore its potential will be better realized when it is integrated with experimental neuroscience. Neuroscience can complement psychiatric genetic epidemiology by specifying the more proximal role of nervous system reactivity in the gene environment interaction"

(Caspi and Moffitt 2006, 584). Genetic, environmental, and neurological causal models of psychiatric disorders are compatible and integrable. They are not reducible or unifiable.

Clearly, not all cases of multiple models of a single phenomenon stand in the same relationship to each other. They might stand in a reduction relationship, if in fact all the causal content of one can be represented, without loss, by the other. They might be in competition, where at best only one should be retained and the other eliminated. The focus of this chapter is on the possibility and value of integrative relationships, where models are neither reductive nor incompatible. In the next section, I turn to protein structure prediction to illustrate compatible perspectival models and the ways in which integrative strategies can produce not knowledge of more features but more accurate knowledge about the same phenomenon.

3 Multiple Perspectives, Multiple Goals

There are many ways to characterize the multiple perspectives displayed by predictive models of functional protein structure. One can crudely distinguish between physical, chemical, and biological perspectives. These are loosely correlated with different contexts in which proteins are studied: in silico, in vitro, and in vivo, respectively. The three perspectives each provide partial and not completely overlapping accounts of the phenomenon. Learning about protein structure from the physics perspective, considering the basic atomic components of proteins and forces acting on them, will inform, but not determine, what is detected from an investigation of the protein's chemical structure. Knowing the chemical details, in turn, informs, but does not completely specify, biogenesis, interaction, and the biological functions of the macromolecule. At the beginning of advances in the study of protein structure in the 1950s, it was believed that a reductive approach, that is, ab initio modeling the thermodynamic features of atomic components of the amino acid sequence of a protein, would be sufficient for predicting the tertiary, biologically functional structure. As Francis Crick proposed, "it is of course possible that there is a special mechanism for folding up the chain, but the more likely hypothesis is that the folding is simply a function of the order of the amino acids" (Crick 1958, 144). This view contains two inferential steps: that protein sequence contains all the information necessary to determine structure and that protein structure is sufficient to determine its function (Berg, Tymoczko, and Stryer 2002; Dill, Ozkan, Weikl, Choder, and Voelz 2007).

However, developments in protein science have revealed a more complex story. Anfinsen's discovery (Sela, White, and Anfinsen 1957) of the spontaneous refolding of a denatured protein in vitro seemed to support the reductionist hopes that nothing more than the interatomic interactions of the atoms making up the primary structure of a protein were required to determine the thermodynamically lowest energy state in a particular environment and that would allow inferences to biological function. Indeed, it was on the basis of Anfinsen's discovery in support of the "thermodynamic hypothesis" (Anfinsen 1973) that he was awarded a Nobel Prize in 1972. Yet in 1969 Levinthal offered a thought experiment that generated a "paradox" for this hypothesis. The number of possible configurations a protein could acquire is astronomical (10^{143}). If configurations were sampled sequentially, it would take longer than the age of the universe for a protein to find its energetically minimum structure. However, proteins fold sometimes in milliseconds. Levinthal claimed that protein structure is not a simple derivation from the physics of the component parts finding their native state: "if the final folded state turned out to be the one of lowest configurational energy, it would be a consequence of biological evolution and not of physical chemistry" (Levinthal 1968, 44).

Some scientists today still hope to find the holy grail of a reductive algorithm that will predict a protein's functional structure. But that hope has not been realized (Mitchell and Gronenborn 2017). Instead there is a proliferation of models, methods, and representations aiming to provide the means to predict protein structure. Rather than viewing the physics models of proteins reductively, I suggest we consider them perspectivally.

How can one model protein transformation from a string of amino acids into a functional three-dimensional structure? On the basis of what features can its structure be predicted? These questions have been investigated by looking at the phenomenon from different perspectives, but the answer cannot be obtained from any one alone. The *functional* structure is not a consequence of the atomic arrangements in the amino acid sequence alone, nor of the hydrophobic and hydrophilic responses of the atoms of a protein with the surrounding molecules in its environment alone, nor of the complex interactions with other proteins that are implicated in the biogenesis of complex proteins in the messy environment of a cell. All three perspectives are needed.

The physics perspective targets the thermodynamic features of folded and unfolded proteins, that is, their mean free energy, and the kinetics of change from denatured through intermediate states to the native state. Coarse-grained and all-atom approaches are used to calculate atomic interactions and to simulate energy landscapes. The unique amino acid string specifies the atomic components of the protein, and the energy content of all possible configurations is computationally sampled to determine which structure possesses the lowest free energy. Thus, the physics perspective investigates protein folding in silico. In answer to Levinthal's paradox, if the energy surface on which the atoms of a protein move is appropriately biased, then even a stochastic search can lead to native structures in realistic times (computationally, as well as in vitro or in vivo). For some proteins, intermediate states act as local minima between the denatured and native states, constituting kinetic traps that can stall the search or prohibit reaching the stable native structure (Bryngelson, Onuchic, Socci, and Wolynes 1995). But where does information about the bias of the energy surface come from? It is not exclusively found in the atomic components of the amino acid sequence; it may be the result of activities of other molecules in the protein's cellular context.

One track through the chemistry perspective targets the detectable three-dimensional structure of a protein by experimentally manipulating an actual protein (chemically or thermally denaturing it, for example) and allowing it to fold in a simplified environment in vitro. The solution conditions of the protein, like temperature, pH, and salt concentration, can be varied to be more or less similar to what would be found in a living cell. The structure is detected by means of x-ray crystallography or nuclear magnetic resonance spectroscopy, with the relative positions or distances between the atoms in the protein computed from x-ray diffraction patterns or spectra shifts. The chemical, experimental perspective that tracks protein interactions in vitro will be discussed in detail in section 4.

The biology perspective studies the protein in the cells in which it is born, functions, and dies—that is, in vivo. The cellular habitat is densely populated, consisting of tens of millions of molecules, and it is in this environment that small and large proteins alike have to fold into their native three-dimensional structures to realize their biological function. Fluorescence microscopy can be used to "see" inside the cell, but it cannot resolve anything at very fine detail or in a very fast time frame. However, evidence of interactions of an unfolded protein with other proteins, chaperones and chaperonins for example, has been detected in the process of folding (Hartl and Hayer-Hartl 2009).

When proteins are made on the ribosome, the leading portion of the polypeptide chain is produced prior to the completion of the anterior portion. This affords the opportunity of the initial portion of the sequence to bind aberrantly to other molecules in the cell, prior to the formation of the anterior region that may be required for interactions that lead to correct folding. Some molecules, called chaperones, bind to the amino terminus of the growing polypeptide chain, stabilizing it in a partially folded configuration (safe from any interfering molecules) until synthesis of the polypeptide is completed. Release of the posterior sequence from the ribosome and detachment of the chaperones then permits anterior and posterior portions of the chain to bind and thus for the protein to fold into its functional three-dimensional conformation.

Each of the three perspectives investigates the problem of predicting protein structure. But they differ in the features targeted for study, the habitat in which the system is studied, and the methods and the representations used to describe what is known about the system. No single approach targets all the features that are relevant to predicting structure. If no representational model is complete, then why should one adopt a particular perspective with its target features, methods, and representations? It is here that pragmatic interests enter. What the investigator wants to *do* provides the source of criteria for judging representational model adequacy. Different models can each correctly describe the same complex system and yet not be reducible to a single representation from a single perspective. Empirical confirmation warrants correctness, while pragmatic concerns decide adequacy. For example, if one wants to determine the relative energetic stability of different conformations of a protein under different conditions, without regard to the specific set of conditions that are found in a living cell, then the physics perspective may deliver results that are both correct and adequate. However, if the goal is to use knowledge of the structure to explore therapies for misfolding diseases, then it will be necessary to go beyond the boundaries of the physics perspective (Karplus 1997).

Ab initio physics models provide insight into the physical constraints and dynamics protein conformations must meet. When the final folded conformation of a protein is thermodynamically stable, that is, has the least free energy, then to move to that conformation from the initial unfolded state is theoretically spontaneous. Chemical processes follow paths toward least free energy. Thus in ideal circumstances, predicting the functional structure of a protein would be finding the thermodynamically stable conformation for that string of amino acids.

However, kinetic factors will decide whether the thermodynamically stable state occurs in fact, under specific environmental conditions, as well as which pathway or pathways through the energy landscape the process of folding is likely to take, and whether it will trapped in a local minima or reach the least energetic conformation. These features of protein folding are targeted by chemical and biological perspectives. The conditions under which proteins fold might be under the control of other molecules, the chaperones, which are part of the cellular environment, the in vivo habitat that is not modeled by either physics in silico or chemistry in vitro methods. "Protein folding occurs in vivo in the environment quite unlike that under experimental, in vitro conditions. A large fraction of newly synthesized protein chains do not fold spontaneously but are assisted by molecular chaperones" (Kmiecik and Kolinski 2011, 10283). This study showed that periodic distortion of the polypeptide chains by chaperone interactions can promote rapid folding and lead to a decrease in folding temperature, changing the conditions under which thermodynamic stability would be defined. It also demonstrated how chaperone interactions can prevent kinetically trapped conformations, thus providing a mechanism for reaching a pathway to the least energy conformation.

In summary, the relationships among the three perspectives for studying protein structure is not one of reduction but rather one of integration. Each provides a partial grasp of the phenomenon, and each requires input and ongoing engagement with the other perspectives, especially if the aims are real-world complex, like finding therapies for misfolding diseases like Alzheimer's and Parkinson's disease. This is how the partiality of perspectival models that leave some factors out (interactions with other molecules in the example above) can be informed and corrected by the perspectives that target those very features. But what if two perspectives target the very same features of the phenomenon and predict what appear to be incompatible models?

4 Multiple Models, One Target

How do we accommodate multiple models that disagree or diverge? I will consider the situation in which the goal is the same—characterizing the atomic structure of a protein—but the methods used to do this are different and the representational models they generate of the same protein diverge. Prima facie, this appears to be a case of conflict. But I will argue that for some divergent models perspectivism offers new insights on how they can be used together.[4]

The plurality of scientific perspectives of protein structure can be used to increase our knowledge of nature, not only by filling out more features, as in the case of chaperones above, but by correcting the systematic biases of different methods. Scientific models derived from different experimental devices, for example, can work in this way. Different experimental protocols are used to detect the structure of a protein. They target different features of the protein and appeal to different background assumptions and theories in order to interpret the detected signals and generate a predictive model of the protein. When different experimental results agree, this is taken as a defense of the reality of the result (Norton 2000). But what should be inferred in the case where the results disagree? I will consider a case where two protocols, x-ray crystallography and nuclear magnetic resonance (NMR) spectroscopy, engaging with the same protein, yield *different* predictions of the protein's tertiary structure.

X-ray crystallography and NMR spectroscopy are the most used and trusted experimental methods to generate predictive models at the atomic level of resolution of protein structure. Indeed, the results of these experiments are taken as arbiters in judging the success of competing algorithms for predicting protein structure from sequence (Moult, Pederson, Judson, Fidelis 1995). Ab initio approaches were discussed above; however, the most successful algorithms for predicting the structure of large proteins from their amino acid sequences are semi-empirical, constructively using data from x-ray and NMR experiments in their development (Mitchell and Gronenborn 2017). When x-ray and NMR target the same protein, the structures that result can diverge. If we describe those representations as propositional claims about the actual protein in the world, then we are faced with claiming that at most one of the representational models is true. Perspectivism allows a different resolution to divergence, one that more accurately captures scientific practice.

X-ray crystallography and NMR spectroscopy involve setting up experiments to causally interact and measure features of a protein that are accessible to their different instruments. For x-ray experiments, a protein is crystallized and then a beam of incident x-rays is diffracted by the electron clouds of the atoms of the protein into many specific directions. The angles and intensities of the diffracted beams are measured to produce a three-dimensional electron density map. From this, the mean positions of the atoms in the crystal can be inferred. In contrast, for NMR the target protein is prepared in solution and placed inside a very large magnet. By so doing, the normal spin of the nuclei of the different atoms in the protein is realigned by the magnetic field of the instrument. A radio pulse is introduced to disrupt this new equilibrium and the return to equilibrium is tracked, revealing information about the effects not just of the experimental magnet but of other nearby atoms. Measurements provide a map of how the different atoms are chemically linked—thus how close they are spatially.

In short, the x-ray perspective detects electron cloud reflectance, while the NRM detects nuclear magnetic changes. Each has a partial perspective on the target phenomenon—namely the atomic tertiary structure of the protein. X-ray experiments allow an inference to the relative positions of atoms, and NMR experiments permits inference to the relative distance of atoms. Clearly the scientific models of the phenomenon are each partial and perspectival, and they generate different predictive representations by using different methods to interact with different target features. When the two experimental protocols are operating as they should and yet deliver different predictions, which one is correct? I suggest that just as in the case of sensory modalities vision and audition discussed above, the answer is both. Each experimental modality correctly detects the features it targets and correctly represents the protein from that perspective.

In addition, in the case of x-ray and NMR experimental models of proteins, there is also a correlate to multisensory integration, namely joint refinement. Joint refinement is a method of mutual error correction that recognizes the different systematic biases that are present in each of the experimental approaches. A joint refinement procedure investigates the compatibility of the two data generated by the two methods and, when possible, attributes differences to the instrumental biases. For example, proteins that are crystallized are in a different physical state from those in solution in the NMR experiments, and that difference may account for some of the divergence in the predicted structures. Other sources of divergence include different degrees of error in the data retrieved or in the ranges of uncertainty in the inferential algorithms (Carlon et al. 2016). When the known systematic biases have been resolved, then the data from the two experiments can be used to predict a protein structure that is more accurate than what could be obtained by either method alone. The blind spots of x-ray crystallography and NMR cannot be removed, but when they are system relative, then they can be exposed by the mutual analysis of joint refinement. "Joint structural refinement using both NMR and x-ray data provides a method to obtain a more reliable structural model, which may disclose additional relevant information on its functional mechanisms" (Carlon et al. 2016, 1601).

Notice that there is no third perspective from which to judge the accuracy of either the x-ray or NMR perspective, the view from nowhere of the objective protein. What scientists have in order to judge the accuracy of any representative model are the data. When you have data from one perspective, it provides justification for a predictive structure of a protein. When you have data from a plurality of perspectives, it can provide stronger justification for a predictive structure. Joint refinement is a form of integration that is pluralism preserving. At the end of joint refinement, there is not global unification of two methods into one or reduction or elimination of two methods to a single perspective. Just as in the case of sensory integration, each of the modalities, vision and audition, can be improved to make better discriminations and increase precision. However, you cannot teach the eyes to hear or the ears to see. The information afforded by a phenomenon to the different modalities when jointly integrated can yield more accurate information than any one sense could ever yield. In both cases, the preservation of pluralism is the means to increase our knowledge of nature.

5 Conclusion

The goal of this chapter was to illustrate the relation between scientific perspectivism and a view I have advocated over the years called "integrative pluralism." Contrary to viewing a plurality of perspectives as exposing the temporary inadequacies of sciences en route to either unification or reduction, multiple perspectives are best seen as enhancing our ability to explain or predict phenomena. This is well illustrated by the case of protein folding, where a plurality of perspectives is seen in the plurality of predictive methods that can yield different representations of the structure of a protein. Yet it is precisely via this exercise of integrating different perspectives—while retaining their unique features—that new insights about the real nature of proteins can be gained.

Notes

1. Robustness in biological systems is one such property that requires appeal to system-level properties. Robustness is the ability of a structure to maintain a system-level function despite internal or external perturbation. It is the dynamic network of relations rather than the properties of the relata that explain this property (Kitano 2004; Mitchell 2008).
2. This is akin to what Cartwright (1983) means by how laws "lie," in that they fail to map exactly onto their domains, containing both abstractions and idealizations. See also Teller's (2001) rejection of the perfect model.
3. See Longino (2013) for a contemporary defense of a form of incommensurability.

4. This is a partial answer to Morrison (2011) who claims, "it isn't clear how perspectivism can help us solve the problem of interpreting the information that inconsistent models provide" (343).

References

Anfinsen, C. B. 1973. "Principles That Govern the Folding of Protein Chains." *Science* 181(4096): 223–230.

Bechtel, W., and Richardson, R. C. 2010. *Discovering Complexity*. Cambridge, MA: MIT Press.

Berg, J. M., Tymoczko, J. L., and Stryer, L. 2002. *Biochemistry*, 5th ed. New York: W. H. Freeman.

Bryngelson, J. D., Onuchic, J. N., Socci, N. D., and Wolynes, P. G. 1995. "Funnels, Pathways, and the Energy Landscape of Protein Folding: A Synthesis." *Proteins* 21(3): 167–195.

Carlon, A., Ravera, E., Hennig, J., Parigi, G., Sattler, M., and Luchinat, C. 2016. "Improved Accuracy From Joint X-ray and NMR Refinement of a Protein-RNA Complex Structure." *Journal of the American Chemical Society* 138: 1601–1610.

Cartwright, N. 1983. *How the Laws of Physics Lie*. Oxford: Oxford University Press.

Caspi, A., and Moffitt, T. E. 2006. "Gene-environment Interactions in Psychiatry: Joining Forces With Neuroscience." *Nature Reviews Neuroscience* 7: 583–590.

Chirimuuta, M. 2016. "Vision, Perspectivism, and Haptic Realism." *Philosophy of Science* 83(5): 746–756.

Crick, F.H.C. 1958. "On Protein Synthesis." *Symposia of the Society for Experimental Biology* 12: 138–163.

Dill, K. A., Ozkan, S. B., Weikl, T. R., Choder, J. D., and Voelz, V. A. 2007. "The Protein Folding Problem: When Will It Be Solved?" *Current Opinion in Structural Biology* 17(3): 342–346.

Feyerabend, P. 1970. *Against Method*. Minneapolis: University of Minnesota Press.

Giere, R. N. 2006a. *Scientific Perspectivism*. Chicago: University of Chicago Press.

Giere, R. N. 2006b. "Perspectival Pluralism." In *Scientific Pluralism*, edited by Kellert, S. H., Longino, H. E., and Waters, C. K., 26–41. Minneapolis: University of Minnesota Press.

Giere, R. N. 2010. "An Agent-based Conception of Models and Scientific Representation." *Synthese* 172(2): 269–281.

Hartl, F. U., and Hayer-Hartl, M. 2009. "Converging Concepts of Protein Folding in Vitro and in Vivo." *Nature Structural & Molecular Biology* 16(6): 574–581.

Karplus, M. 1997. "The Levinthal Paradox: Yesterday and Today." *Folding and Design* 2(Supplement 1): S69–S75.

Kitano, H. 2004. "Biological Robustness." *Nature Reviews Genetics* 5: 826–837.

Kmiecik, S., and Kolinski, A. 2011. "Simulation of Chaperonin Effect on Protein Folding: A Shift From Nucleation-Condensation to Framework Mechanism." *Journal of the American Chemical Society* 133: 10283–10289.

Levinthal, C. 1968. "Are There Pathways for Protein Folding?" *Journal de Chimie Physique* 65(1): 44–45.

Longino, H. E. 2013. *Studying Human Behavior*. Chicago: University of Chicago Press.

Mitchell, S. D. 2000. "Dimensions of Scientific Law." *Philosophy of Science* 67(2): 242–265.

Mitchell, S. D. 2003. *Biological Complexity and Integrative Pluralism.* Cambridge: Cambridge University Press.

Mitchell, S. D. 2008. "Exporting Causal Knowledge in Evolutionary and Developmental Biology." *Philosophy of Science* 75(5): 697–706.

Mitchell, S. D. 2009. *Unsimple Truths: Science, Complexity, and Policy.* Chicago: University of Chicago Press.

Mitchell, S. D., and Gronenborn, A. M. 2017. "After Fifty Years, Why Are Protein X-ray Crystallographers Still in Business?" *The British Journal for the Philosophy of Science* 68(3): 703–723.

Morrison, M. 2011. "One Phenomenon, Many Models: Inconsistency and Complementarity." *Studies in History and Philosophy of Science Part A* 42(2): 342–351.

Moult, J., Pederson, J. T., Judson, R., and Fidelis, K. 1995. "A Large-scale Experiment to Assess Protein Structure Prediction Methods." *Proteins* 23(3): ii–iv.

Norton, J. D. 2000. "How We Know About Electrons." In *After Popper, Kuhn and Feyerabend; Recent Issues in Theories of Scientific Method,* edited by Nola, R., and Sankey, H., 67–97. Dordrecht: Kluwer.

Perini, L. 2012. "Image Interpretation: Bridging the Gap From Mechanically Produced Image to Representation." *International Studies in Philosophy of Science* 26(2): 153–170.

Robinson, G. E., Grozinger, C. M., and Whitfield, C. W. 2005. "Sociogenomics: Social Life in Molecular Terms." *Nature Reviews Genetics* 6: 257–270.

Sela, M., White, F. H., Jr., and Anfinsen, C. B. 1957. "Reductive Cleavage of Disulfide Bridges in Ribonucleas." *Science* 125(3250): 691–692.

Stein, B. E., and Stanford, T. R. 2008. "Multisensory Integration: Current Issues From the Perspective of the Single Neuron." *Nature Reviews Neuroscience* 9: 255–266.

Suárez, M. 2010. "Scientific Representation." *Philosophy Compass* 5(1): 91–101.

Tabery, J. 2014. *Beyond Versus: The Struggle to Understand the Interaction of Nature and Nurture.* Cambridge, MA: MIT Press.

Teller, P. 2001. "Twilight of the Perfect Model Model." *Erkenntnis* 55(3): 393–415.

van Fraassen, B. C. 2008. *Scientific Representation: Paradoxes of Perspective.* Oxford: Oxford University Press.

Contributors

Hasok Chang is Hans Rausing Professor of History and Philosophy of Science in the Department of History and Philosophy of Science at the University of Cambridge and the British Academy's Wolfson Research Professor in 2017–2020. He is the author of *Inventing Temperature* (Oxford University Press, 2004) and *Is Water H₂O?* (Springer, 2012), and has extensively written in the area of history and philosophy of the physical sciences.

Mazviita Chirimuuta is Associate Professor in the Department of History and Philosophy of Science at the University of Pittsburgh. She specializes on the philosophy of neuroscience and philosophy of perception, and is the author of *Outside Color* (MIT Press, 2015).

David Danks is L. L. Thurstone Professor of Philosophy and Psychology in the Department of Philosophy at Carnegie Mellon University. His research falls within the intersection of philosophy, cognitive science, and machine learning. He is the author of *Unifying the Mind* (MIT Press, 2014) and co-editor, with Emiliano Ippoliti, of *Building Theories* (Springer, 2018).

Melinda Bonnie Fagan is Sterling McMurrin Associate Professor in the Department of Philosophy at the University of Utah. Her research focuses on experimental practice in biology, explanation, and philosophical conceptions of objectivity and evidence. She is the author of *Philosophy of Stem Cell Biology* (Palgrave Macmillan, 2013).

Michela Massimi is Professor of Philosophy of Science at the University of Edinburgh. Her research interests are in the philosophy of science, the history and philosophy of modern physics, and Kant's philosophy of nature. She is the author of *Pauli's Exclusion Principle* (Cambridge University Press, 2005) and co-editor of *Kant and the Laws of Nature* (Cambridge University Press, 2017).

Casey D. McCoy is a Postdoc at Stockholm University. He was previously Teaching Fellow in Philosophy of Science and Postdoctoral Research

Fellow on the ERC-funded "Perspectival Realism" project at the University of Edinburgh. His research is in the philosophy of physics and the general philosophy of science.

Sandra D. Mitchell is Distinguished Professor in the Department of History and Philosophy of Science at the University of Pittsburgh. Her research centers on scientific explanations of complex behavior and the representation of multilevel, multicomponent complex systems. She is the author of *Biological Complexity and Integrative Pluralism* (Cambridge University Press, 2003), *Komplexitäten* (Suhrkamp, 2008), and *Unsimple Truths* (University of Chicago Press, 2009).

Anya Plutynski is Associate Professor in the Department of Philosophy at Washington University in St. Louis. Her research interests are in the history and philosophy of medicine, general philosophy of science, and biomedical ethics. She is the author of *Explaining Cancer* (Oxford University Press, 2018) and co-editor of *The Routledge Handbook to the Philosophy of Biodiversity* (Routledge, 2017) and *A Companion to the Philosophy of Biology* (John Wiley & Sons, 2008).

Collin Rice is Assistant Professor in the Department of Philosophy at Bryn Mawr College and, since 2013, an Associate Scholar of the Center for Philosophy of Science at the University of Pittsburgh. He works primarily in the philosophy of science, the philosophy of biology, and the philosophy of mind.

Juha Saatsi is Associate Professor in Philosophy at the University of Leeds. Much of his research concerns scientific realism and explanation, especially in relation to the history and philosophy of physics. He is the editor of *The Routledge Handbook of Scientific Realism* (Routledge, 2017) and co-editor of *Explanation Beyond Causation* (Oxford University Press, 2018).

Paul Teller is Professor Emeritus in the Department of Philosophy of the University of California Davis, and has been a member of the faculty there since 1990. His research spans the philosophy of physics, metaphysics, logic, and the philosophy of science. He is the author of *A Modern Formal Logic Primer* (Prentice Hall, 1989) and *An Interpretive Introduction to Quantum Field Theory* (Princeton University Press, 2005).

J. E. Wolff is Lecturer in the Department of Philosophy of King's College London (from September 2019 in the School of Philosophy, Psychology and Language Sciences at University of Edinburgh). She was a Humboldt Fellow in the Munich Center for Mathematical Philosophy at Ludwig Maximilian University Munich (2015–2018). Her research is in the philosophy of science and metaphysics.

Index

Note: Page numbers in italics indicate figures and in bold indicate tables on the corresponding pages.